IMAGES
of America

THE COLUMBIA RIVER

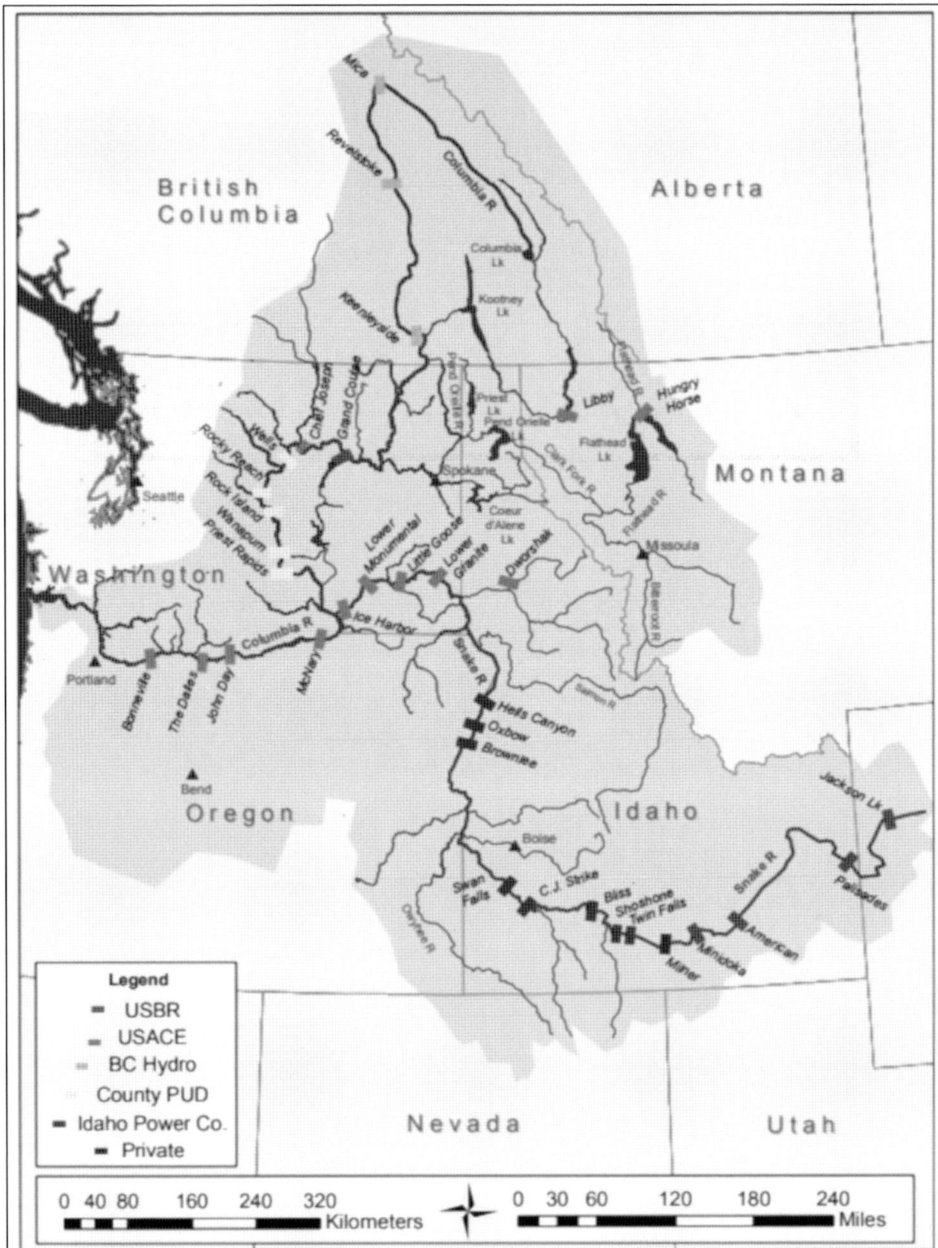

The Columbia River Basin drains 256,000 square miles, an area the size of France, and includes parts of seven states—Washington, Oregon, Idaho, Montana, Nevada, Utah, and Wyoming—and the province of British Columbia. The river begins at Columbia Lake, British Columbia, at an elevation of 2,651 feet, and flows 1,248 miles to the Pacific Ocean. This map shows the major hydropower and flood control dams in the basin. (Courtesy Eastern Washington University.)

On the Cover: This September 17, 1942, photograph shows the danger, difficulty, ingenuity, and joy of fishing at Celilo Falls, which helped sustain Columbia River tribes for thousands of years. Fifteen years after this photograph, the reservoir rising behind The Dalles Dam flooded the falls. (Courtesy John Klatt, Old Oregon Photos.)

IMAGES
of America

THE COLUMBIA RIVER

John A. Harrison

ARCADIA
PUBLISHING

ISBN 978-1-4671-0768-6

Published by Arcadia Publishing
Charleston, South Carolina

Printed in the United States of America

Library of Congress Control Number: 2021948394

For all general information, please contact Arcadia Publishing:
Telephone 843-853-2070
Fax 843-853-0044
E-mail sales@arcadiapublishing.com
For customer service and orders:
Toll-Free 1-888-313-2665

Visit us on the Internet at www.arcadiapublishing.com

To Dawn, who shares my passion for the extraordinary Columbia, and for our grandchildren Tess, Ron, and June.

CONTENTS

Acknowledgments

I want to acknowledge the following people for their contributions of time, information, photographs, support, and cheerfulness: Irene Martin, author and historian, Skamokawa, Washington; Libby Burke, archivist, Bonneville Power Administration, Portland, Oregon; J. Patrick Barry, Vancouver, Washington, US Army Corps of Engineers (retired); Nicholas Simoes, visitor center manager, Grand Coulee Dam, Washington; Jim Lichatowich, biologist, historian, and author, Columbia City, Oregon; Blaine Hardin, author, Seattle, Washington; John Klatt, Old Oregon Photos, West Linn, Oregon; Eileen Delehanty Pearkes, author and historian, Nelson, British Columbia; Guy Moura, Cultural Resources and History, Colville Confederated Tribes; Scott Daniels, Oregon Historical Society; Kyle Kusch, Arrow Lakes Historical Society, Nakusp, British Columbia; Dr. William L. Lang, historian, author, and professor emeritus, Portland State University; Dr. Allan Scholz, historian and professor, Eastern Washington University; Eric Schrepel, Gillian Charles, and John Fazio, my colleagues at the Northwest Power and Conservation Council, Portland; and Stacia Bannerman and Erin Vosgien, Arcadia Publishing.

Introduction

The Columbia is the fifth-largest river by volume in North America. Its huge annual runoff, averaging 192 million acre-feet, pushes a freshwater plume up to 50 miles wide, 150 miles long, and 60 feet deep into the Pacific Ocean. It is a river that drains parts of seven states and the province of British Columbia.

The Columbia is known primarily for two things: hydropower and salmon—more precisely, anadromous fish, which are those born in freshwater, migrate to the ocean as juveniles, and then return to the streams of their birth as adults to spawn. The Columbia became a great hydropower river by virtue of its steep descent—roughly two feet per mile for 1,200 miles—and the solid rock canyon, good for anchoring large dams, through which it flows for much of that distance. It was a great anadromous fish river because of its abundant clear, cold water, thanks to snowmelt runoff from several mountain ranges.

Historically, annual salmon and steelhead runs numbered at least 6 to 8 million—some estimates are more than 20 million. But the dams took a toll on anadromous fish by making passage difficult and, in some cases, blocking passage altogether. The impacts of dams, together with other factors like overfishing and habitat destruction, drove down the annual runs. Salmon and steelhead runs today rarely top 1.5 million fish. Dams without fish passage eliminated access to more than 40 percent of the historic spawning and rearing habitat for anadromous fish. At the same time, Columbia River Basin dams provide nearly half of the region's electricity and also make inland navigation, flood control, irrigation, and recreation possible.

The Columbia, first hypothesized in the 17th century by European geographers and cartographers as the Great River of the West, existed in these hopeful minds long before it was actually discovered by maritime explorers and merchants. It was hoped the river might provide an ice-free passage from the Atlantic to the Pacific. Thus, it was envisioned as a river of commerce, and its first European and American explorers, arriving on the Pacific coast in the 18th century and having almost accidentally discovered the high value of furs like sea otter and beaver in China, envisioned the Columbia as a source of commerce and profit.

Two early explorers of the Columbia extolled its potential for settlement and commerce. In the estuary with Robert Gray in 1792, fifth mate John Boit envisioned "a fine place to set up a factory." Near the Canadian headwaters in 1883, railroad engineer Sandford Fleming wondered "how soon will busy crowds of workmen take possession of these solitudes, and the steam whistle echo and re-echo where now all is silent?"

Thomas Jefferson envisioned the Columbia as part of a trading route across North America to the Pacific. Inland fur-trading routes utilized the river and its tributaries; sea-going vessels increasingly explored the Pacific coast and the lower Columbia in the late 1700s. Columbia River tribes traded salmon from the river with tribes from the interior. It was inevitable that as river commerce grew, river traffic would grow as well. Sail power soon gave way to steam power; river traffic grew rapidly in both the number of commercial vessels and the economic importance of the transportation industry. Steadily, the Columbia River country became more populated and the river became an important artery of fish and commerce.

The first non-Indian visitors to the Columbia noted its size and the importance of salmon to the native peoples. In his book *The Journals of Lewis and Clark*, historian Bernard DeVoto wrote that as the Corps of Discovery journeyed down the Clearwater and Snake Rivers toward the Columbia, "everywhere the party saw evidence of the salmon economy, around which the life of these river tribes was organized: weirs, spears, nets, [made of fiber], caches of dried fish, etc." In his journal for October 17, 1805, Clark wrote about the Columbia near the present-day location of Pasco and Kennewick, Washington: "The number of dead Salmon on the Shores & floating in the river is incredible." Similarly, in June 1811, Canadian explorer David Thompson witnessed Indians fishing at Kettle Falls several hundred miles farther upriver and observed hundreds of salmon leaping over the falls. He wrote, "The salmon are about from 15 to 25 to 30 pounds weight here, well tasted, but have lost all their fat, retaining still all their meat. Their flesh is red and they are extremely well made." He was intrigued by the importance of salmon to the tribes, the rituals they had regarding salmon (for example, like a modern-day conservation practice, a certain number were allowed to pass upriver before fishing began), and the fact that the fish had empty stomachs, meaning they had not eaten on their long journey to spawn.

The Columbia is the Northwest's river, and its story is the story of the Northwest. Author Tim Egan once wrote that the Northwest is "any place a salmon can go." Today, perhaps more precisely, the Northwest is any place a salmon could go before dams blocked their passage. But the implication is clear and accurate. The Northwest is a place where the environment is the economy, and where the Columbia River touches on virtually every aspect of life—dams that enable electricity, navigation, flood control, irrigation, recreation, and fish passage, and a culturally and economically important in-river fishery for Indians and non-Indians. The story of the Columbia is the story of an evolution from a vast wilderness waterway to an industrial machine – an "organic machine," as historian Richard White has called it.

It is the story of "the Big River," a common, if loose, translation of its name in the languages of several indigenous peoples whose homelands include the Columbia: Wimahl in the language of the Upper Chinook Tribe, who lived along the river from the mouth of the Cowlitz River to The Dalles; Nch I'Wana in the language of the Sahaptin-speaking people of the mid-Columbia; Mixqaqas akinmituk, ("River in the Chickadee Territory") in the Kutenai language of the Ktunaxa people who live in the Columbia headwaters region of British Columbia; the mythical, west-flowing River Ourigan to French Canadian fur traders in the 18th century; "River of the West" in 19th-century American literature; and ultimately, Columbia's River, for the name—or part of it—of the ship (*Columbia Rediviva*) American fur trader Robert Gray sailed across its treacherous bar in May 1792, the first recorded crossing.

Indeed, in many ways historical and contemporary, the Columbia is one of the world's great rivers.

One

Creation and Discovery of the Big River

There was a beginning to the Columbia, 90 to 150 million years ago. Over time, the ancestral river carved its way down valleys, across plains, and through the Cascade Mountains, as seen in this view looking west near Hood River, Oregon. (Author's collection.)

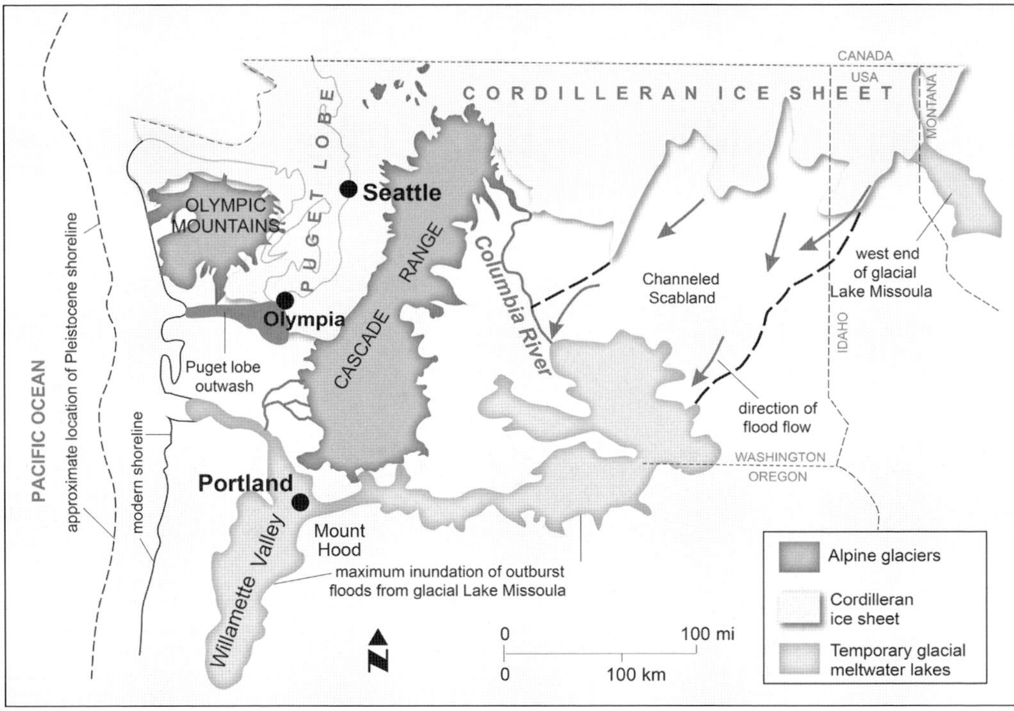

About 12,000 to 15,000 years ago, a series of cataclysmic floods caused by repeated failures of an ice dam in what is now western Montana rampaged down the Columbia River drainage, ripping topsoil from hundreds of square miles of eastern Washington and helping dig the Columbia River Gorge through the Cascade Mountains, where the floodwaters were as deep as 900 feet. Evidence of the floods was first described in 1919 by geologist J. Harlen Bretz. Other geologists did not believe him, but eventually, he was proved right. Ice carried giant boulders, known as glacial erratics, into places in eastern Washington and northern Oregon, where they settled as the ice melted. The erratic at left rests on a hilltop near McMinnville, Oregon, southwest of Portland, 150 feet above the valley. (Above, courtesy Washington Department of Natural Resources; left, author's collection.)

As this 1950 photograph shows, floodwaters unleashed by repeated failures of the ice dam holding back Glacial Lake Missoula exposed and then rushed over a cliff in central Washington, digging a canyon 3.5 miles wide with a drop of 400 feet. Today, the place is called Dry Falls. There is no river, but there is a series of spring-fed lakes. Over time, Dry Falls became a popular tourist destination. Nearby, floodwaters roaring across the landscape carved another deep canyon that became known as the Grand Coulee. Today, both of these deep canyons are reminders of the basin's violent geologic history. (Courtesy Washington State Archives.)

The creation story of the Columbia can be told in two ways: the geologic narrative of volcanism, basalt, floods, and ice, and the narrative of a spirit creating a world for humans and animals. The oral tradition common to Columbia River tribes is that the world was water and that a spirit reached down and scooped mud to make the land. Some of the mud the spirit piled high, like 12,360-foot Mount Adams, or Pahto. The wise trickster Coyote fought a battle with the giant beaver Wishpoosh to open mountains that blocked salmon from the river and its people. The slashing of the beaver's tail formed the Columbia River Gorge. Pieces of the defeated beaver became humans. Remnants of a dam that blocked salmon are the rocks of Celilo Falls. (Above, courtesy Washington State Archives; below, courtesy Old Oregon Photos.)

The first humans to live along the Columbia River migrated between 12,000 and 30,000 years ago. In July 1996, the remains of a 9,300-year-old human were found buried in the Columbia River shore near Kennewick, Washington. Area tribes named the skeletal remains "the Ancient One." DNA matched people from several of those tribes. After a legal battle, the bones were awarded to the tribes, who buried them in a secret location. In the 1950s, bones and artifacts of a similar age were discovered, remarkably well preserved, in a cave near the confluence of the nearby Palouse and Snake Rivers. The site was excavated in the late 1960s, but work could not be completed before the area was flooded by the reservoir rising behind Lower Monumental Dam. (Above, courtesy Brittney Tatchell, Smithsonian Institution; below, courtesy Washington State University Libraries.)

While the first humans to settle in what is now the Columbia River Basin came by land, the next visitors arrived by sea. Perhaps the earliest recorded voyage to pass by the mouth of the Columbia took place in the fifth century AD. The monk Hwui Shan left China and traveled around the rim of the Pacific. A narrative was later included in official state records. In spring 1778, Capt. James Cook, the famed British explorer, missed the mouth of the Columbia altogether. On May 11, 1792, the American fur trader Robert Gray entered the river, the first American or European to do so, and named it after his ship, *Columbia Rediviva*, depicted below in a squall. (Above, courtesy Northwest Power and Conservation Council; below, courtesy Oregon Historical Society.)

Gray explored the river 36 miles inland. Thus, by maritime convention, the Columbia became an American river. Twelve years later, Pres. Thomas Jefferson sent two explorers, Meriwether Lewis and William Clark, west to explore the Missouri River to its headwaters and then the connection, if any, to the Columbia. Lewis and Clark spent the winter of 1805–1806 at a fort they built near present-day Astoria, Fort Clatsop, and then began their homeward journey east up the river the following spring. The fort has been reconstructed and today is the centerpiece of a national historical park. (Courtesy National Park Service.)

The earliest European and American visitors to the Northwest extolled its potential for commerce, primarily in furs, fish, and lumber. Robert Gray's crew recognized this importance in 1792, as had Cook's crew before them, and others even earlier. As these explorers discovered, the skins of marine mammals, mainly sea otters, brought extraordinary prices in Canton, where they were traded for items including tea, spices, and fabric. About 200 miles up the Columbia from the ocean, Lewis and Clark witnessed fishing at Celilo Falls in April 1806, with Clark noting in his journal, "We observed many stacks of fish remaining untouched on either side of the river. This is the great mart of all this country. Ten different tribes visit those people for the purpose of purchasing their fish." (Above, courtesy Oregon Historical Society; left, courtesy McCord Museum, Montreal, M21417.275.)

The North West Company, a Canadian firm, had been exploring the region west of the Rocky Mountains for furs since 1807 when one of its employees, David Thompson, crossed the Continental Divide and discovered the main river, the first to do so in British Canada. Thompson also established the trading post Kootenai House at the northern end of Lake Windermere, a short distance downriver from the headwaters lake. In 1811, he established Boat Encampment on the Columbia between the mouths of the Wood and Canoe Rivers. For more than 50 years, Boat Encampment was an important transfer point for goods being transported between Montreal and the Pacific coast. The site was flooded behind Mica Dam in 1973. (Above, courtesy Library and Archives Canada/Sir Henry James Warre fonds/c026348k; below, courtesy Cliff White Lens of Time Northwest Project.)

In 1811, Thompson became the first European to travel the length of the Columbia, although not completely on the river. Beginning in the headwaters region of the Columbia near the Continental Divide, he traveled southwest along Columbia River tributaries until reaching Spokane House, his trading post on the Spokane River. From there, he followed an Indian trail north to the confluence of the Colville and Columbia Rivers, where he arrived on June 19. A short distance above the mouth of the Colville, he witnessed Indians fishing at Kettle Falls of the Columbia, which he named Ilthkoyape ("basket" in the native language) Falls, and the Indians who fished there Ilthkoyape Indians. The Indians knew the falls as Swah'netk'qhu, which translates as "noisy-sounding water." Kettle Falls was a centuries-old, vibrant fishery. "Kettle" referred to depressions worn in the rocks of the falls. Local Indian tribes—today, the Skoyelpi and Sinixt tribes of the Confederated Colville Tribes—caught salmon with hoop nets and in baskets and also traded fish with visiting tribes. (Courtesy Washington State Archives.)

At about the same time, New York merchant John Jacob Astor, who had made a fortune in the East Coast fur trade, sought to expand his empire to the Pacific through a succession of separate companies that would be established across the country and eventually joined. One of these, which he incorporated in 1810, was the Pacific Fur Company. To establish a trading post on the Pacific, Astor dispatched two parties to the mouth of the river in September 1810, one by sea, in his ship *Tonquin*, and the other by land. The ship arrived in the Columbia River estuary in late March 1811, where eight men died as the ship attempted, and finally succeeded in, crossing the bar. The crew built a stockade, which they named Fort Astoria, pictured here in 1845. (Courtesy Oregon Historical Society.)

Life was hard at Fort Astoria, as it was chronically undersupplied. In 1813, the North West Company bought the fort and renamed it Fort George, in honor of Great Britain's King George III, operating it for many years. Meanwhile, the Hudson's Bay Company moved its Columbia Department headquarters inland and built Fort Vancouver on the north shore of the river at the present-day location of Vancouver, Washington, in 1824. The original Fort Vancouver was built on a bluff above the river, but in 1828–1829, it was relocated about a mile to the west. The fort is reconstructed on that site today. (Above, courtesy Washington State University Libraries; below, author's collection.)

By the 1840s, people were flooding into the Northwest, many on the 2,000-mile-long Oregon Trail. The Dalles, Oregon, became an important decision point for emigrants, who either loaded their belongings on a raft and floated down the Columbia to the Willamette River and then up to Oregon City, or took the Barlow Trail around Mount Hood and down to Oregon City. Either was risky but particularly the passage through the Cascade Rapids, pictured here. Many lost their possessions and their lives. Others chose to settle around The Dalles rather than risk the perilous trip to Oregon City. (Above, courtesy Washington State Archives; below, courtesy Old Oregon Photos.)

Inland fur-trading routes utilized the river and its tributaries; sea-going vessels increasingly explored the lower Columbia waterway beginning in the late 1700s. It was inevitable that as river commerce grew, river traffic would grow as well. Sail power soon gave way to steam power. The first steam-powered vessel on the Columbia was, fittingly, a Hudson's Bay Company ship, the *Beaver*, arriving in 1836. Pictured here in Victoria, British Columbia, the *Beaver* could operate with either sails or steam, and under steam power, its boilers consumed up to 40 cords of wood per day. (Courtesy Vancouver Maritime Museum.)

Two

Furs, Fish, Mining, Timber, and New Beginnings

Salmon was a primary food of Columbia River tribes for millennia, but with the influx of immigrants, particularly from Scandinavia, salmon became the basis of an industry that grew rapidly along the lower Columbia River, supplying canneries like this one at Cathlamet, Washington, photographed about 1900. Frank Warren, who built the Warren Packing Company cannery in 1869, died in the sinking of the *Titanic* in 1912. (Courtesy Old Oregon Photos.)

Missionaries and fur traders moved into the sovereign territories of Columbia River tribes. Missions were built by both Catholics and Protestants. While many have faded into history, one that is remembered to this day as a national historic site is the Waiilatpu Mission of the missionary and doctor Marcus Whitman. Waiilatpu was about nine miles west of present-day Walla Walla, Washington. As a missionary, Whitman failed to convert any Cayuse; as a doctor, the Cayuse blamed him for an epidemic of measles. On November 29, 1847, the Cayuse attacked, killing Whitman, his wife Narcissa, and 12 others. The mission had become a focal point for the growing alienation, anger, and distrust the Cayuse felt over the steady influx of Oregon Trail pioneers. (Above, courtesy National Park Service; below, author's collection.)

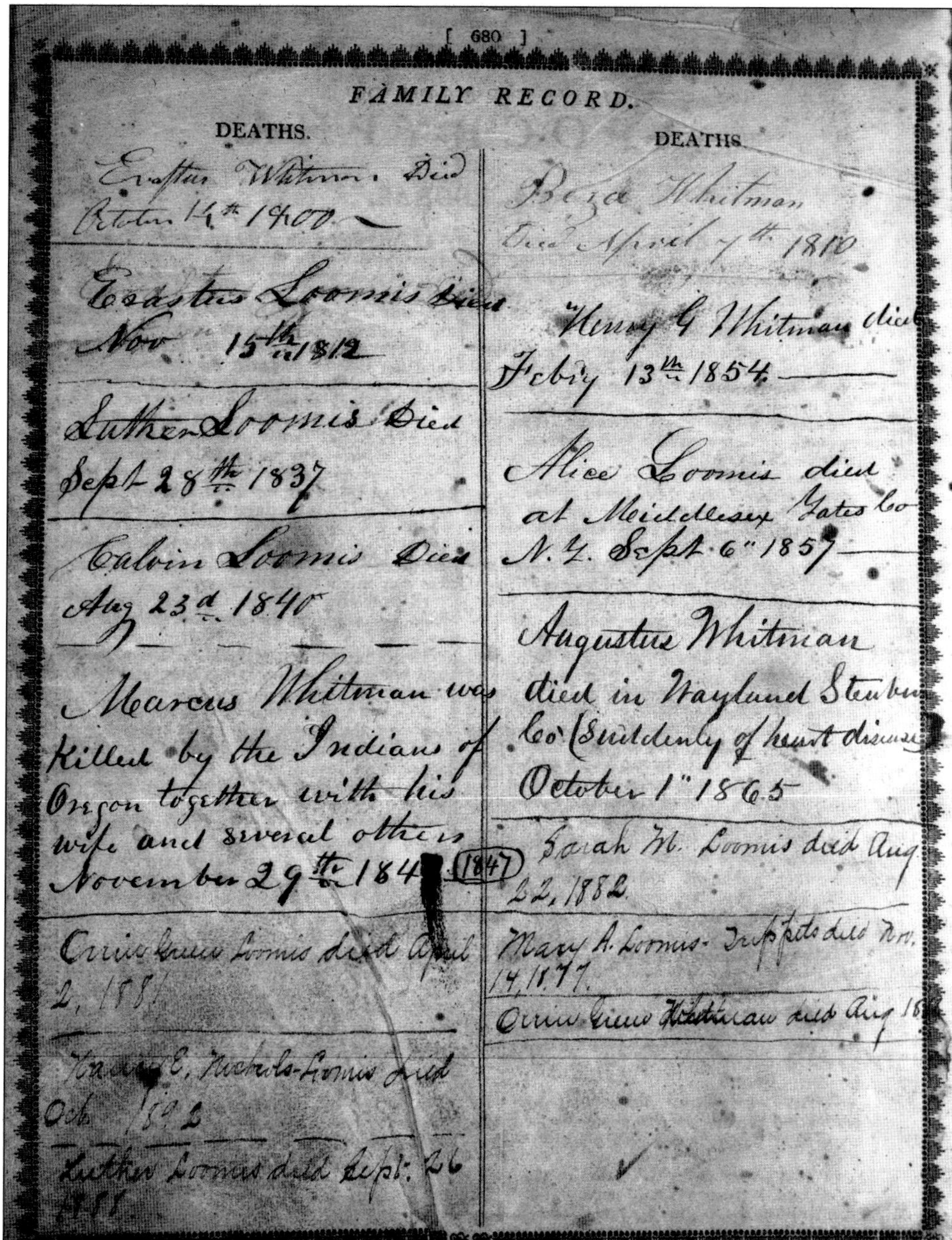

[680]

FAMILY RECORD.

DEATHS.	DEATHS.
Erastus Whitman. Died October 14th 1800	Beza Whitman Died April 7th 1810
Erastus Loomis Died Nov 15th 1812	Henry G Whitman died Feby 13th 1854
Luther Loomis Died Sept 28th 1837	Alice Loomis died at Middlesex Yates Co N.Y. Sept 6" 1857
Calvin Loomis Died Aug 23d 1840	Augustus Whitman died in Wayland Steuben Co (Suddenly of heart disease) October 1" 1865
Marcus Whitman was Killed by the Indians of Oregon together with his wife and several others November 29th 184[illegible] (1847)	Sarah M. Loomis died Aug 22, 1882
Orrin Green Loomis died April 2, 1881	Mary A. Loomis-[illegible] died Nov. 14, 1877
[illegible] E. Nichols-Loomis died Oct 1892	Orrin Green Whitman died Aug 18[illegible]
Luther Loomis died Sept. 26 1881	

The notes in the Bible of Marcus Whitman's mother record the deaths at the mission. The handwritten text reads, "Marcus Whitman was killed by the Indians of Oregon together with his wife and several others, November 29th, 1847." Over time, the legacy of Marcus and Narcissa Whitman changed as more facts were uncovered and published about their mission and work. The Whitman name is affiliated with streets, places, and even a college in Walla Walla. The missionaries were long heralded as heroes for trying to convert the native peoples. But in more recent decades, the narrative has shifted to a story of two idealistic Christians who failed to understand the native way of life and whose relationship with the local Indians deteriorated over time, ending with their murders. (Courtesy Washington State University Libraries.)

Narcissa Whitman, the wife of missionary Marcus Whitman, described Fort Vancouver in the 1840s as a place so alive and bustling that it was, to her, "the New York City of the Pacific." The Hudson's Bay Company's supervisor at the fort, John McLoughlin, was "the chief figure of Northwest history," in the words of historians Dorothy Johansen and Charles Gates. McLoughlin considered the land north of the river British, despite the policy of joint occupation with the United States in place since 1818. When Oregon Trail pioneers arrived, he welcomed them hospitably—and turned them south. The first railroad along the river opened in 1851 on the Washington shore, hauling freight and passengers past the Cascades Rapids. It had wooden rails and one wagon, pulled by a mule. (Both, courtesy Old Oregon Photos.)

One of the most famous shipwrecks among the more than 200 at the mouth of the Columbia River is that of the USS *Shark*, a Navy warship that visited in July 1846. Its commander, Lt. Neil M. Howison, had been ordered to survey the Oregon Territory. In September, while attempting to leave, the *Shark* foundered and sank while crossing the bar. The crew survived but was stranded near the mouth of the river. They built a log house and named the place Sharksville. There, in October 1846, Howison received news from an inbound merchant ship that the United States and Great Britain had signed the Treaty of Oregon on June 15, and the territory was now American. Howison then became the first person to fly the American flag—rescued from the *Shark*—over the now American-only Oregon territory. The memory of the *Shark* remains today in the name of Cannon Beach, Oregon. There, three pieces of its artillery were found on the beach. All are now in museums. (Courtesy US Navy, Naval History & Heritage Command.)

The fur trade was the first dominant business in the Northwest, with competition among American, Canadian, and British companies. The biggest and most powerful of these was the Hudson's Bay Company, which built or acquired a string of trading posts across the interior Columbia River Basin. The British company built Fort Colville on the southeast bank of the Columbia at Kettle Falls in 1826, five years after merging with its rival North West Company of British Canada. Fort Colville quickly grew in importance in the company's operations. The farm was particularly successful. Wheat initially did poorly but later thrived, as did vegetables and livestock. Resolution of the international border in 1846—Fort Colville now was in America—and the steady decline of the fur trade, coupled with the unrest of Indian wars in the 1850s, led the company to gradually shift operations north into Canada. The US Army established its own Fort Colville in 1859, about 14 miles south of the Hudson's Bay Company's post. (Courtesy Washington State Archives.)

Beavers were important in Columbia River history, particularly in the early 1800s. Their fur was thick and water-repellent, and they could be caught easily. Their skins proved to be a valuable trade commodity and were particularly sought-after in China, where British and American merchants traded beaver skins for exotic goods like tea and silk. Beaver fur coats and hats were popular too in America and Europe. Indians also relied on beavers for clothing, particularly in the inland areas where winter temperatures were colder than on the Pacific coast. But competition was ferocious. Beginning about 1820, the Hudson's Bay Company set out to exterminate beavers in the Columbia River country in an effort to drive out the competition. "It was a conscious policy of destruction designed to forestall US competition," historian Robert Bunting wrote. (Courtesy Mcgill Library, HJJdGJrRQ84.)

The popularity of beaver top hats, like the one pictured above at the restored Fort Vancouver, waned after 1825. In that year, which was the same year the Hudson's Bay Company dedicated Fort Vancouver, a silk top hat made its fashion debut in Paris. Suddenly, beaver was out and silk was in, at least for top hats, and the market for the skins collected by the Bay Company began to decline. There are few photographs of beaver pelts from the late 1800s, as there were by then few beavers. Here, a Colville couple poses with a river otter pelt, probably at Fort Colville about 1895. (Above, author's collection; left, courtesy Old Oregon Photos.)

In 1818, the North West Company established a fur-trading fort at the confluence of the Walla Walla and Columbia Rivers and named it Fort Nez Perces, after the local Indian tribe. When the Hudson's Bay Company bought the North West Company in 1821, the name changed to Fort Walla Walla. After the attack on the Whitman Mission in 1847, relations between settlers and tribes in the area steadily worsened, and in 1855, the US Army recommended settlers in the area leave, as further hostilities were possible. In fact, a battle was fought nearby that year. Today, the location of the fort is underwater in the pool behind McNary Dam. (Above, Washington State Archives; below, author's collection.)

Near present-day Walla Walla, Washington, in June 1855, the US government convened a council of area Indian tribes, including the Cayuse, Umatilla, Nez Perce, Palouse, Yakama, and Walla Walla. Led by Washington Territory governor Isaac I. Stevens and Oregon Indian Affairs superintendent Joel Palmer, the government wanted the tribes to agree to move to reservations. This was a foreign concept to the semi-nomadic tribes, who attended the council to voice their views about their land, their heritage, and their beliefs. Palmer and Stevens brushed aside these concerns and insisted that the tribes agree to move to reservations. In short, the government threatened the tribes, and after about a week of intense discussions, the tribes acquiesced, reserving their right to fish at streams on and adjacent to the reservations and "at all other usual and accustomed stations in common with citizens of the United States." Those words in the Walla Walla treaty and others negotiated with other Northwest tribes proved critical in future court battles over fishing rights. This Gustav Sohon painting depicts Indians at the Walla Walla treaty council. (Courtesy Washington State Historical Society.)

Christian missionaries (both Catholic and Protestant) and fur traders were among the first non-indigenous people to move into the Columbia River Basin, and they were soon followed by many seeking other kinds of opportunities. As the California gold rush of the 1840s slowed, miners pushed north. Gold was discovered near Fort Colville and farther north in British Columbia in the 1850s, bringing miners who trespassed over newly created Indian reservations, prompting hostilities, notably a short war with the Yakama Nation. The US Army closed eastern Washington Territory to further settlement in 1856 in response to the war, but miners were exempted. A gold rush to central Idaho and eastern Oregon between 1861 and 1870 brought waves of people up the Columbia to try their luck and settle the interior. In 1861 alone, 10,500 men came up the river. Steamboat traffic on the Columbia and Snake Rivers increased significantly, hauling freight and passengers to mining districts on both sides of the border. Pictured here from left to right are the steamers *Rossland*, *Kootenay*, and *Minto* on the upper Columbia at Nakusp, British Columbia. (Courtesy Arrow Lakes Historical Society.)

Sandon, British Columbia, in the valley of the Columbia River tributary Slocan River, was typical of many towns that were founded around gold and silver mines and that eventually died as the bounty of ore declined. Silver was discovered in Sandon Creek in 1891, and silver mines in that area were the richest in the province for about 10 years after that. The town was named for John Sandon, a California man who staked the first mining claims in the area. Two railroads raced to be the first to serve booming Sandon, arriving in 1895. Sandon was the first fully electrified town in the province, thanks to a hydroelectric plant that serviced a mine. The town grew fast and at its peak had about 5,000 residents. Over time, the town declined along with the output of nearby mines. During World War II, Sandon hosted an internment camp for Japanese Canadians from the Pacific coast. Large-scale mining continued until 1953, and today, Sandon is a ghost town and tourist attraction. (Courtesy Arrow Lakes Historical Society.)

While Celilo Falls was arguably the largest Indian fishery and trading place on the river, Kettle Falls, about 500 miles upriver from Celilo, was the highest waterfall (about 50 feet) on the Columbia and also had a historic tribal fishery and trading center. There, Indians fished primarily with dip nets, but also with woven baskets placed at the base of the falls to catch fish as they fell back while attempting to pass. Kettle Falls was not only an important Indian fishery, but also an attraction for non-Indian visitors, as the 1887 photograph below shows. (Above, courtesy Washington State Archives; below, courtesy Northwest Room, Spokane Public Library.)

For millennia at Kettle Falls, local Indian bands caught salmon and steelhead for their own use and also to trade for goods brought by Indians who traveled to the fishery from as far away as present-day Montana. Historical accounts suggest the fishery at Kettle Falls might have been larger than the one at Celilo. Between 1,000 and 2,000 Indian fishers regularly used the site during salmon runs, particularly in the summer. In July 1811, British explorer David Thompson, director of the North West Company's Columbia Department, viewed fishing at Kettle Falls and wrote it had "by far the largest population of Indians, all heavily dependent upon the salmon." In 1841, Charles Wilkes, commander of the US Exploring Expedition, observed Indians trapping salmon in baskets at the falls. He wrote that the Indians frequently caught as many as 900 salmon in 24 hours. (Courtesy Washington State Archives.)

At about the same time as the mid-1800s gold rush to the interior of the Columbia River Basin, other entrepreneurs recognized the commercial potential of the river's abundant salmon. From humble beginnings, salmon canning grew to become the major industry of the Columbia River. At the industry's peak, in the 1880s, there were 39 canneries on the Columbia. Brothers George W. and William Hume brought salmon canning to the Columbia River in 1866. They had opened a cannery on the Sacramento River in California in 1864 but, disappointed by the runs, moved to the Columbia with their partner Andrew Hapgood and opened a cannery on the Washington shore about 50 miles inland at a place they called Eagle Cliff. The first year, they packed 4,000 cases, 48 one-pound cans to the case, by hand. But as the salmon were steadily overfished, the canneries declined. The last one closed in 1980. This interior view of the Eagle Cliff cannery is from about 1920. (Courtesy Columbia River Maritime Museum.)

In his 1893 book about the salmon canning industry, Robert (R.D.) Hume lamented the decline of salmon that was evident by then, 27 years after his brothers built the first Columbia River cannery. Ironically, the decline was largely the result of poor regulation of the fishing that supported the industry that made him rich. He called the salmon industry "the new El Dorado." By the early 1870s, the Hume family was operating several Columbia River salmon canneries. To lower production costs, George Hume brought workers from China. He was the first to employ Chinese workers in American canneries. Hume preferred Chinese workers because they were efficient and hardworking and would accept low pay. Most of the fishing for the canneries was done by local Indians; the Chinese were not allowed to fish. (Above, courtesy Irene Martin; below, courtesy Oregon Historical Society.)

Many emigrants from the East Coast of the United States were attracted by the potential of salmon fishing on the Columbia. The Hume brothers, for example, were from Maine. As well, men and families with fishing experience arrived on the lower Columbia from Scandinavia, forming distinct communities of Swedes, Norwegians, and Finns. There was a large community of Finnish immigrants in Astoria and a distinctly Norwegian community of fishing families on Puget Island in the Columbia about 60 miles inland. US government reports about Columbia River fisheries informed fishers in Europe and Scandinavia about the Columbia's potential, as did articles in newspapers. Many people were also eager to leave difficult political conditions in late 19th-century Europe. Some communities, like Skamokawa, Washington, grew up around fishing and logging. In a town with no roads, women rowed their children to school, as seen in this 1880s painting. (Courtesy Irene Martin.)

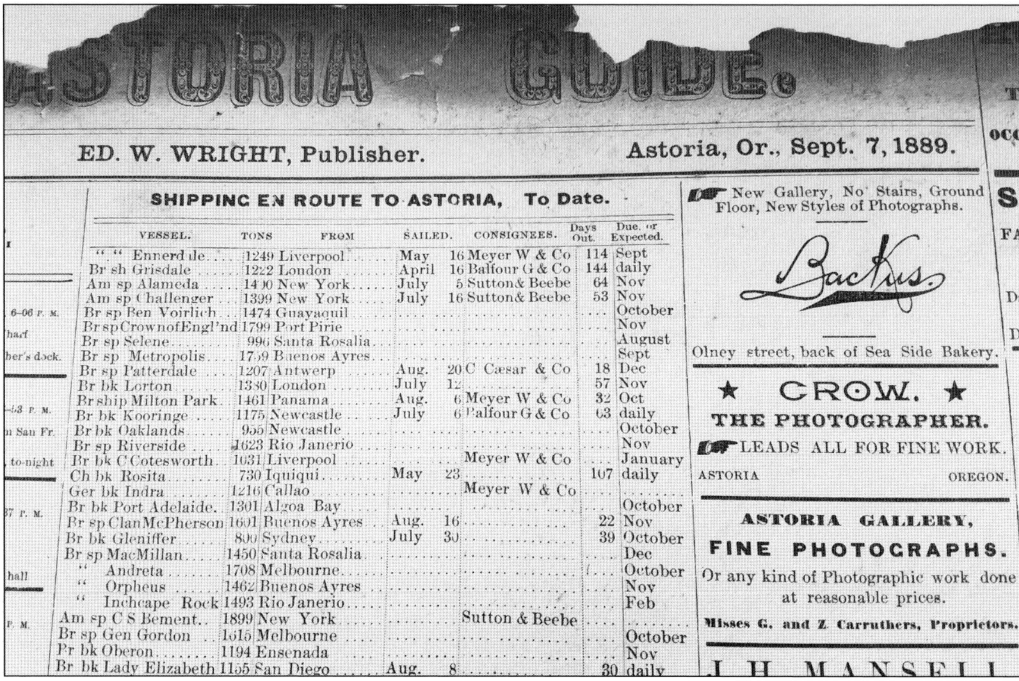

ED. W. WRIGHT, Publisher. Astoria, Or., Sept. 7, 1889.

SHIPPING EN ROUTE TO ASTORIA, To Date.

VESSEL.	TONS	FROM	SAILED.		CONSIGNEES.	Days Out.	Due. or Expected.
" " Ennerdale	1249	Liverpool	May	16	Meyer W & Co	114	Sept
Br sh Grisdale	1222	London	April	16	Balfour G & Co	144	daily
Am sp Alameda	1400	New York	July	5	Sutton & Beebe	64	Nov
Am sp Challenger	1399	New York	July	16	Sutton & Beebe	53	Nov
Br sp Ben Voirlich	1474	Guayaquil					October
Br sp Crown of Engl'nd	1799	Port Pirie					Nov
Br sp Selene	996	Santa Rosalia					August
Br sp Metropolis	1759	Buenos Ayres					Sept
Br sp Patterdale	1207	Antwerp	Aug.	20	C Cæsar & Co	18	Dec
Br bk Lorton	1330	London	July	12		57	Nov
Br ship Milton Park	1461	Panama	Aug.	6	Meyer W & Co	32	Oct
Br bk Kooringe	1175	Newcastle	July	6	Balfour G & Co	63	daily
Br bk Oaklands	955	Newcastle					October
Br sp Riverside	1623	Rio Janerio					Nov
Br bk C Cotesworth	1031	Liverpool			Meyer W & Co		January
Ch bk Rosita	730	Iquiqui	May	23		107	daily
Ger bk Indra	1216	Callao			Meyer W & Co		
Br bk Port Adelaide.	1301	Algoa Bay					October
Br sp Clan McPherson	1601	Buenos Ayres	Aug.	16		22	Nov
Br bk Gleniffer	800	Sydney	July	30		39	October
Br sp MacMillan	1450	Santa Rosalia					Dec
" Andreta	1708	Melbourne					October
" Orpheus	1462	Buenos Ayres					Nov
" Inchcape Rock	1493	Rio Janerio					Feb
Am sp C S Bement.	1899	New York			Sutton & Beebe		
Br sp Gen Gordon	1515	Melbourne					October
Br bk Oberon	1194	Ensenada					Nov
Br bk Lady Elizabeth	1155	San Diego	Aug.	8		30	daily

By the late 1890s, Astoria, the oldest non-indigenous community on the Columbia, was booming. The 1890 population was 6,184; by 1900, it was 8,351, an increase of 35 percent. As this clip from the September 7, 1889, issue of the *Astoria Guide* shows, ships were arriving from all over the world. Steamers left daily for cities upriver. The passenger fare to Portland, 98 miles upriver, was \$2. Astoria even had an office of the White Star Line, the Liverpool company that owned the *Titanic*. (Both, courtesy Irene Martin.)

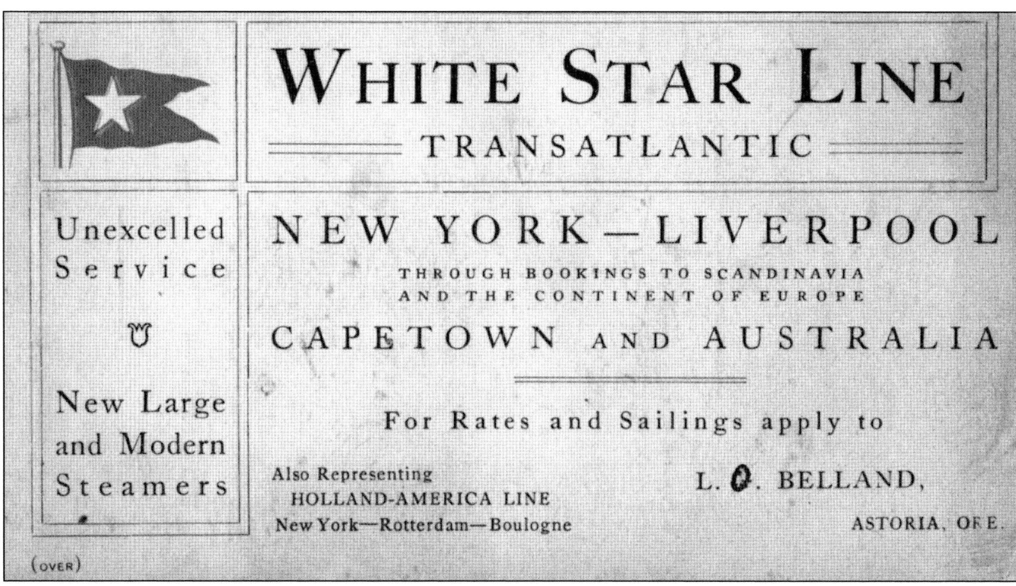

Three

Boom Times on the River

In 1892, the *Coulee City* [Washington] *News* reported on a proposal to irrigate the Big Bend country of central Washington state with water diverted from a large dam on the Columbia River. This probably is the first published proposal for a dam at Grand Coulee. Here, the Keller and Republic Stage depot at the town of Keller shows just how dry the area was. (Courtesy Bureau of Reclamation.)

This 1894 photograph by James Bratt captures a detailed view of Astoria. The city's maritime economy focused on shipping and fishing, but railroads and logging were also important. For a time, Astoria was the second-largest city in Oregon behind Portland. By the late 1890s, Columbia River fishing and canning, primarily focused on salmon, was the largest industry in the Northwest. Between Astoria and the Cascades Rapids of the Columbia, about 130 miles upstream, some 30 canneries were operating, and about 2,500 fishers were employed, mostly by the canneries using their gear. The huge log on a single flatcar below shows the enormous size of the trees around Astoria, seen here at Clatsop Junction about 1910. (Above, courtesy Old Oregon Photos; below, courtesy Washington State Archives.)

Salmon, managed by tribes in fisheries that had existed for thousands of years, were so abundant that they sometimes flooded the floors of canneries as workers collected them for processing from wholly unregulated fisheries. The only identification for this c. 1915 photograph is "a Columbia River salmon cannery," but the scene would have been common at many of the canneries along the lower Columbia River. (Courtesy Washington State Archives.)

By 1890, largely as a result of the completion of transcontinental railroads, the combined population of Idaho, Washington, and Oregon jumped from about 251,000 in 1880 to 705,000. The population of British Columbia was growing as well, thanks to the completion of Canada's transcontinental railway as well as mining, logging, and farming. Above, a passenger train crew and passengers pose with the locomotive, probably at Rosebery, British Columbia, on Slocan Lake. Below, the Oregon Trunk Line bridge crosses the Columbia near Celilo Falls east of The Dalles, creating a freight and passenger link between Washington and Oregon. (Above, courtesy Arrow Lakes Historical Society; below, courtesy Oregon Historical Society.)

Around the turn of the 20th century, there were some three dozen canneries along about the first 140 miles of the Columbia. Demand for canned salmon was high during World War I, as much of the nation's beef was reserved for the troops. Demand remained stable, both in the United States and in Europe, through the 1920s, when canned salmon was a staple food in many households, easily acquired and prepared. Columbia River salmon was shipped around the world. Between 1891 and 1895, Columbia River salmon canneries packed an average of 486,000 cases (48 one-pound cans per case) each year, and between 1889 and 1920, the average annual harvest of Chinook salmon from the lower Columbia River was 25 million pounds. These women are applying labels to cans of salmon at the J.G. Megler and Company Cannery at Brookfield, Washington, about 40 miles inland. (Courtesy Oregon Historical Society.)

Over time, there were hundreds of canned salmon brands, each with its own distinct label. Some are works of art, and are now collector's items. Irene Martin of Skamokawa, Washington, a well-known author, Columbia River historian, and retired gillnet fisher with her husband, Kent, collects labels and said these are among her favorites. She said the Gladiola Brand label (above) is one of the most beautiful, and the Beacon Brand label has a unique feature—the shining beacon that Lady Liberty is holding is a salmon. The Pillar Rock and Boss labels below show sail-powered gillnet boats and also demonstrate the tight-knit community of gillnetters. Unique or hard-to-find labels might sell for as much as $600, and many sell for $20–$50. (Both, courtesy Irene Martin.)

In his 1949 book *The Columbia: Powerhouse of the West*, author Murray Morgan described horse seining on the lower Columbia: "From the highway you can see the draft horses moving shoulder deep in the green-gray water as they haul the seine into position, then plodding ashore and dragging out of the river a brown net flecked with silver or, just as often, clogged with weeds and driftwood." Hauling in the salmon was a spectator event, at times. Below, a group watches the beaching of nets full of salmon on Sand Island near Chinook and Ilwaco, Washington, in 1903. (Both, courtesy Irene Martin.)

Farther upriver, salmon fishing also was a way of life, a sport, and a commercial business. Indians had fished at Celilo Falls, about 200 miles inland, for millennia. Their usual method, as the men in the photograph above show, was to dip hoop nets on long poles into the river. Before dams blocked their migration, salmon and steelhead were caught as far upriver as the headwaters by both First Nations fishers and others. In the 1916 photograph at left, a man poses with a large salmon, its spawning readiness evident by its dark color, near the community of Brisco on the Columbia River about 55 miles north of Columbia Lake and more than 1,100 miles from the ocean. He shot the fish from the shore with a rifle. (Above, courtesy Washington State Archives; left, courtesy Bob Ede.)

For Columbia River tribes, salmon and steelhead are a reliable and culturally important food source. Prior to dams and reservoir flooding, tribes caught the fish at traditional sites along the river, such as Celilo Falls and Kettle Falls, from early spring through the fall. While an important mainstay, salmon were not the only food or even the major one for most tribes. Ethnohistorical surveys suggest fish provided 30–40 percent of caloric needs for mid-Columbia and Plateau tribes, with other foods such as roots, berries, ungulates, and waterfowl making up the rest. In this 1902 photograph, a Wishram woman prepares salmon in a traditional manner at Celilo. Salmon was eaten fresh and also dried and packed into bricks, often with berries, for consumption in the dormant season. In treaties with the US government in 1855, Indians of the mid-Columbia River reserved the right to fish and hunt at their "usual and accustomed" places. (Courtesy Old Oregon Photos.)

Fish wheels were an ingenious but highly exploitative type of commercial fishing gear, prominent in the Columbia River Gorge beginning in the 1880s. They were metal mesh basket traps built on wheels that turned by the force of the swift river current, scooping hundreds of tons of fish from the river and dumping them onto barges—sometimes so many fish that their weight sank the barges. In his 1980 memoir *Wheels of Fortune*, Francis Seufert, grandson of the founder of the Seufert Brothers family fish-packing business at The Dalles, called fish wheels "by far the most fascinating gear that was ever invented. There was something about the wheels dipping into the water and coming out with the fish that fascinated people no end." The Seufert business owned a 10-mile stretch of the Oregon side of the river where it located fish wheels, and also a fleet of some 50 gillnet boats operated by hired fishermen. (Courtesy Oregon Historical Society.)

In the late 1800s, when the Columbia River fishery seemed limitless, one person's economic opportunity was another person's economic—and cultural—loss. Historian Richard White comments in his 1995 book about the Columbia, *The Organic Machine*, that fish wheels were built at traditional Indian fishing sites, pushing aside the Indians who had treaty rights to fish dating to 1855. In the mid-1920s, the Oregon Grange initiated a measure banning fish wheels. Public concern was growing over the impacts of commercial gear like fish wheels and the waste of fish when the wheels caught more than the canneries could handle. Salmon and steelhead runs had been declining annually for several years. Oregon voters approved the measure in November 1926. This prompted some wheel owners to move their gear to the Washington side of the river, where wheels were not banned until 1935. (Courtesy Old Oregon Photos.)

Along with salmon, the other great resource of the Columbia River Basin ripe for exploitation was its vast forests. At first, it was a free-for-all—timber was cheap compared to elsewhere in the country. Because of occasionally deep snow, trees were often cut 20 feet or so from the ground, leaving tall stumps. Logs and timber were cut at odd and random lengths with little attention to standard grades and dimensions common in other parts of the United States. In Washington alone, by 1890, the annual cut totaled a billion board feet or more. The influx of immigrants to the Northwest increased the demand for lumber for housing. With the completion of transcontinental railroads in the late 1800s, lumber shipped from Columbia River cities like Portland found eager markets in the Midwest. These men are felling a tree in the lower Columbia area of Washington. (Courtesy Irene Martin.)

Mechanization came slowly to the logging industry. Around the turn of the 20th century, much logging still was done with horses and handsaws, as the 1903 photograph above of men working for the Shafer & Weist Logging Company of Bayview, Washington, shows. Mechanization increased production, and as early as 1904, mills were overproducing lumber compared to demand from consumers. Machines like the steam-powered drag saw pictured below in a Washington logging camp around 1900 greatly increased the volume of timber that could be cut. It was called a drag saw because it only cut on the backstroke. At about this time, public concern began to grow over cutting trees on public land, and so timber companies increasingly began buying vast acreages of forest to supply their mills. (Above, courtesy Irene Martin; below, courtesy Old Oregon Photos.)

Logging was big business, and so were the logs. Above, employees of the Alger Logging Company in southwest Washington pose for a photograph in 1908. Vast amounts of logs and lumber were exported to the Midwest and California. Simon Benson, a lumber company owner in Portland, developed a means of transporting timber south via the ocean by cabling together a million board feet or more of logs at a time. Called the Benson raft—or the cigar raft, for its shape—this was a practical and inexpensive way, compared to transport by railroad, to supply large amounts of timber to California, where Benson had a sawmill in San Diego. Below, a 600-foot cigar raft tied off at Astoria waits for an ocean tugboat. (Above, courtesy Irene Martin; below, courtesy Old Oregon Photos.)

Logs from forests bordering the lower Columbia were shipped around the world in sailing vessels. Above, the ships *Big Bonanza*, left, and *Louis*, right, load lumber at Skamokawa, Washington, on the lower Columbia in 1892. Early pioneers from Europe marveled at the forests, with trees larger than any they had ever seen. Pacific smelt, also known as eulachon, have always been ubiquitous in the lower Columbia, fished by simply scooping them out of the water as they make their annual spring migration up the river. Below, smelt dippers crowd the river near Longview, Washington. (Above, courtesy Irene Martin; below, courtesy Washington State Archives.)

The importance of salmon extended beyond the border into Canada, both for settlers and indigenous people. Grand Coulee Dam, completed at river mile 596 in 1941, blocked salmon and steelhead migration to the upper Columbia. Salmon had long returned in great numbers to lakes and rivers in British Columbia, as shown in this 1908 photograph of Eli Edge and his catch of big salmon (and a duck) at Comaplix, British Columbia, about 900 miles from the ocean. Their size suggests they were summer Chinook, also known as June Hogs for their girth and weight. Upper Columbia summer Chinook were the largest salmon in the Columbia system. Comaplix was a short-lived (1898–1915) sawmill town on the Northeast (Beaton) Arm of Upper Arrow Lake. The town was destroyed by a catastrophic fire and the site was flooded by the Keenleyside Dam reservoir in 1968. (Courtesy Arrow Lakes Historical Society.)

The Columbia River in British Columbia was rich with fish, particularly from spring to fall. Above, Eli Edge and an unidentified friend show off another string of big ones, and two ducks, again at Comaplix around 1910. The c. 1910 photograph below shows that white sturgeon were also taken in the river, this one weighing 161 pounds and caught at Burton, 22 miles south of Nakusp in The Narrows, a 14-mile stretch of wide river that formerly separated Upper and Lower Arrow Lake. Since the completion of Keenleyside Dam in 1968, the two lakes have become one reservoir but are still referred to separately by locals. (Both, courtesy Arrow Lakes Historical Society.)

Five Canadian First Nations recognize the mainstem of the Columbia River as important for their traditional use, an area stretching from what is now the international border at river mile 749 to the headwaters at Columbia Lake, 1,248 miles from the ocean. Above, a man, woman, child, and dog are shown in about 1907–1909. The sturgeon-nose canoe, inspired by the shape of the head of white sturgeon, was a common design among First Nations people. The family are likely Sinixt/Lakes people, whose homeland was along the Columbia River mainstem from Kettle Falls to the Arrow Lakes and Revelstoke, British Columbia. The photograph at left, taken in 1907, shows shelters of the Lakes people near Nakusp on the Upper Arrow Lake. (Both, courtesy Arrow Lakes Historical Society.)

Commercial steamers began operating in the Columbia by the mid-1800s. In 1850, the *Columbia* and the *Lot Whitcomb* went into operation. The 90-foot-long *Columbia*, the first steamer constructed on the river, was built at Astoria and put into service between there and Oregon City. The *Lot Whitcomb*, which was larger, was built at Milwaukie, Oregon, on the Willamette River upstream from Portland. In 1853, the *Eagle* was assembled at the upper end of the Cascades Rapids and put into service between there and The Dalles. The *Colonel Wright*, built in 1859, was the first steamboat launched above The Dalles. The Dalles-Celilo Canal was built between 1905 and 1915 to allow boats to bypass Celilo Falls without having to transfer cargo and passengers at The Dalles. (Above, courtesy Bonneville Power Administration; below, courtesy Library of Congress.)

Meanwhile, other steamboats were being built that would open other parts of the Columbia for water transportation. In 1865, the *Forty Nine*, a steamboat loaded with hopeful miners and their equipment, left Fort Colville for the Big Bend of the Columbia in British Columbia, where gold had been discovered. It was the first steamboat to cross the 49th parallel on the Columbia. The boat was stopped by ice about 240 miles south of the Big Bend. The following year, 1866, the *Forty Nine* tried again and made it as far as Downie Creek, 47 miles north of Revelstoke, a one-way voyage of 270 miles. The one-way fare was $25 per person, and freight was $200 per ton. Pictured are steamers at Nakusp, above, and St. Leon, below, both on the Arrow Lakes, in 1909. (Both, courtesy Arrow Lakes Historical Society.)

Steamboats also operated among the small Columbia River communities in the headwaters region. On May 8, 1886, Capt. Frank Armstrong launched the first steamboat on the upper Columbia River, at Golden, British Columbia. The ship was called the *Dutchess*. A large crowd of First Nations people, likely Ktunaxa, met the ship at Lake Windermere on its maiden voyage upriver; it was the first steamboat they had seen. When the ship ran aground on a salmon-spawning gravel bar, the First Nations people helped pull it free. In 1891, Armstrong incorporated the Upper Columbia Navigation & Tramway Company to haul passengers and freight between the Columbia and Kootenay Rivers. The south-flowing Kootenay passes within a quarter-mile of Columbia Lake, the headwaters of the Columbia, which from there flows north toward Lake Windermere, Golden, and the Big Bend. Here, the steamers *JD Farrell* and *North Star* are tied up at Jennings, Montana, on the Kootenai (spelled with an "i" in the United States). (Courtesy British Columbia Archives.)

One of the most interesting, if not oddest, bits of Columbia River steamboat trivia is the brief but spectacular history of the Baillie-Grohman Canal, which linked the Kootenay River with Columbia Lake. The quarter-mile-long canal was the work of William Adolph Baillie-Grohman, a visionary entrepreneur, wealthy business promoter, and adventurer who wanted to drain floodplain for agriculture at Kootenay Lake, downriver on the US/Canada border. In exchange for receiving land concessions there from the provincial government, he proposed building a canal linking Columbia Lake to the Kootenay River, near what became the community of Canal Flats. He built the canal, but it was used only three times. The third time it was used, it was destroyed. (Left, courtesy Nick Knowles, creative commons license; below, courtesy Fort Steele Historic Town Archives.)

Capt. Frank Armstrong, who built the *Dutchess* at Golden, also built the *North Star* at Jennings, Montana, on the Kootenai River. In July 1902, he ran the *North Star* from Jennings upriver to Baillie-Grohman's canal, more than 200 miles. There, he discovered that the 130-foot ship was too long (by 30 feet) and too wide (by nine inches) for the lock gates. Unperturbed, Armstrong cut the side rails off the ship, burned away the lock gates, and lengthened the lock with dams of sandbags at either end. Then he set a dynamite charge in the north dam, blew it up, and sluiced the big boat into Columbia Lake. The sand-bag dam at the south end kept the *Kootenay* from illegally flowing into the lake. Armstrong then ran the *North Star* down to Golden, where the leading citizens were so impressed they hosted a dinner in his honor. One year later, the boat was seized at Golden for failure to pay the duty for entering Canada, and 10 years after that, the boat was scrapped. (Courtesy Royal British Columbia Museum, A-09279.)

When Captain Armstrong blew out the north end of the canal, he also essentially blew up any hope of ever using the canal again for commercial purposes. For decades, it remained a local curiosity as it slowly deteriorated. In this photograph, men inspect the remains of the canal in 1914. By 1920, when the adventurer and author Lewis R. Freeman visited the site at the beginning of his canoe journey down the Columbia, he saw a collapsing wall of old logs and assumed it was part of the canal foundation. He also saw traces of the canal. These remain visible to this day. (Courtesy Fort Steele Historic Town Archives.)

By the beginning of the 20th century, lumbering and mining were booming industries on the upper Columbia River in British Columbia, attracting immigrants from countries as diverse as Russia and Italy who formed distinct ethnic communities in towns along the river, such as Trail and Castlegar. These places had once been Sinixt villages, fisheries locations, and gathering sites. The discovery of gold on Red Mountain in 1890 brought miners to Trail and nearby Rossland. A large community of Italian immigrants grew in Trail—by one estimate, 20 percent of the population. A benevolent society, the Columbo Lodge, was formed to support Italian families. In 1895, F.A. Heinze of Butte, Montana, built a smelter to handle the Red Mountain ore. In 1898, Heinze sold the smelter to the Canadian Pacific Railway. Soon, it was also processing lead and zinc. In 1906, the smelter, three local mines, and the Rossland Power Company collectively formed the Consolidated Mining and Smelting Company of Canada. The smelter in Trail is pictured here in 1910. (Courtesy Trail Historical Society Museum.)

Today, the smelter, pictured here in 1970, is owned and operated by Teck Resources Ltd. of Vancouver, British Columbia. The smelter continues to dominate the riverfront and skyline of Trail. The Teck facility is one of the largest zinc and lead smelters in the world, producing precious and specialty metals, chemicals, and fertilizer products. Between 1896 and 1995, the smelter discharged millions of tons of waste material into the river and toxic emissions into the air. Discharged waste settled on the riverbed and beaches a short distance downriver in Washington state, poisoning fish and other aquatic organisms. Litigation over Teck's responsibility for polluting a US water body continues. (Courtesy Arrow Lakes Historical Society.)

Immigrant Doukhobors were a deeply religious community of Russian Christians persecuted for rejecting the Orthodox Church and for their refusal to bear arms. In 1898, more than 7,000 emigrated to Canada, where they had been promised free land and exemption from military service. Initially settling in Saskatchewan, the majority came to British Columbia between 1908 and 1912. Their administrative center and a jam factory were at the confluence of the Columbia and Kootenay Rivers, where they supplanted a large Sinixt village, burial ground, and fishery. Their leader, Peter Verigin, also directed the construction of a large, rectangular two-and-a-half-story communal house on the river shore at Trail. Doukhobors were vegetarians known for their prolific fruit and vegetable gardens. Most Doukhobor buildings are gone or abandoned. The flooded riverfront communal house is pictured below in 1914. (Both, courtesy Trail Historical Society.)

By the early 1900s, transportation was evolving quickly with the growth of population and the demands of logging, mining, and agriculture. The 1916 Columbia River Highway, the first paved highway in the Northwest, ran along the south (Oregon) side of the river between Troutdale (14 miles east of Portland) and The Dalles, a distance of 74 miles. The Latourell Falls Bridge is pictured above in 1915. The scenic highway was envisioned by entrepreneur and good-roads promoter Sam Hill and his friend Samuel Lancaster, an engineer. Lancaster located the highway to take careful and dramatic advantage of the scenery and vistas, modeling the road after the Auxenstrasse in Switzerland, with beautifully crafted stonework, cliffside bridges, and dramatic tunnels with windows cut in them. (Above, courtesy Oregon Historical Society; below, Washington State Archives.)

In 1928, Portland publisher Lipschuetz and Katz issued *Oregon's Famous Columbia River Highway*, which describes itself as "a descriptive View Book in Colors." The book contains 35 images documenting the magnificent views and roadside pullouts along the river both upstream and downstream from the Portland/Vancouver area. The book and the promise of amazing scenery likely enticed people from the Portland area to drive out to hike and picnic in the Columbia River Gorge. The Columbia River Highway runs in front of 600-foot Multnomah Falls, about 30 miles east of Portland, seen at right in an image from the book. The falls remain a popular tourist and hiking destination, along with the lodge near the base of the falls. (Both, author's collection.)

MULTNOMAH FALLS FROM THE HIGHWAY.

In the late 1800s, concern for the effects of overharvesting and canning millions of pounds of salmon and steelhead from the Columbia River each year prompted a new idea: raising fish artificially in hatcheries, augmenting their natural spawning. This idea came from no less an expert than Spencer Baird, the commissioner of the US Commission on Fish and Fisheries, in response to a request for advice from the Oregon legislature in 1875. Baird based his recommendation on the apparent success of artificial propagation of salmon in the Sacramento River. Cannery owners, fearing the alternative was regulation of the fishery, raised $25,000 for a hatchery at Clackamas, Oregon. More hatcheries were built, including this one by the Oregon Fish Commission at Bonneville in 1915. By 1922, it was clear that artificial propagation of salmon had not conserved the runs, which continued to decline. Historian and salmon biologist Jim Lichatowich writes in his book *Salmon Without Rivers*: "With absolutely no scientific evidence, Spencer Baird launched a myth and a program that would have profound effects on the salmon." (Courtesy Oregon Historical Society.)

Four

The Once and Future Workhorse River

The years from 1920 to 1933 were a time of intense optimism about the future of the Columbia. In 1920, Washington Highway 14 was completed on the north side of the river, connecting the Portland/Vancouver area with points cast. This 1920s photograph shows a car rounding the corner in the area known as Cape Horn, with a magnificent view of the river 500 feet below. (Courtesy US Army Corps of Engineers.)

In 1915, Washington and Oregon created the Columbia River Compact to resolve the chaos that resulted every year when the two states established and tried to enforce their own commercial fishing seasons and regulations in the lower Columbia River. Congress approved the compact in 1918. Despite these improved harvest policies, a noticeable decline in salmon and steelhead began in 1921. In later years, the river was divided into six fishing zones: five downstream of Bonneville Dam for non-Indian commercial fisheries, and one above Bonneville exclusively for Indian tribes. At left, Tony Canessa poses with an 85-pound Chinook salmon in Astoria in 1925. (Above, courtesy Library of Congress; left, courtesy Columbia River Maritime Museum.)

With the advent of hydroelectric power generation in the late 1800s, attention in the Northwest began to turn to building dams. The Columbia Plateau of central Washington, surrounded on three sides by the river, was also a prime location for large-scale irrigation, if Columbia River water could only be delivered there. Two plans emerged: one backed by private business interests for a canal that would deliver water to the plateau from some upstream location, and another for a large dam at the Grand Coulee that would impound a lake, from which water could be pumped to the plateau. The ensuing fight pitted the dam "pumpers" against the gravity plan "ditchers." In 1921, the Washington Water Power Company of Spokane sought a permit from the Federal Power Commission to build a dam at Kettle Falls, 110 miles upstream from Grand Coulee. A dam there would essentially kill the pumpers' plan, and they protested. The permit was issued the following year, but the dam was never built. (Courtesy Washington State Archives.)

The fight between pumpers and ditchers carried on through the 1920s. The focus was on irrigation, not on the impacts a large dam, either at Kettle Falls or at Grand Coulee, would have on Columbia River fish runs or the native peoples who depended on the fishery culturally, economically, and nutritionally. The tribes were not silent. In January 1921, tribal leaders from Washington and Oregon gathered at the Washington state capitol in Olympia to lobby the legislature for the right to fish at Prosser Falls in the Yakima River, where the federal Reclamation Service had built a large-scale irrigation project beginning in 1905. These and related events in the 1920s presaged the era of big-dam building on the Columbia that would begin within a decade. (Above, courtesy Northwest Room, Spokane Public Library; below, Washington State Archives.)

Protect the Salmon ... VOTE 332 (X) YES

(The Fish-Wheel, Trap, Seine and Gill Net Bill)

This picture shows why the salmon cannot reach their spawning grounds on the upper Columbia River

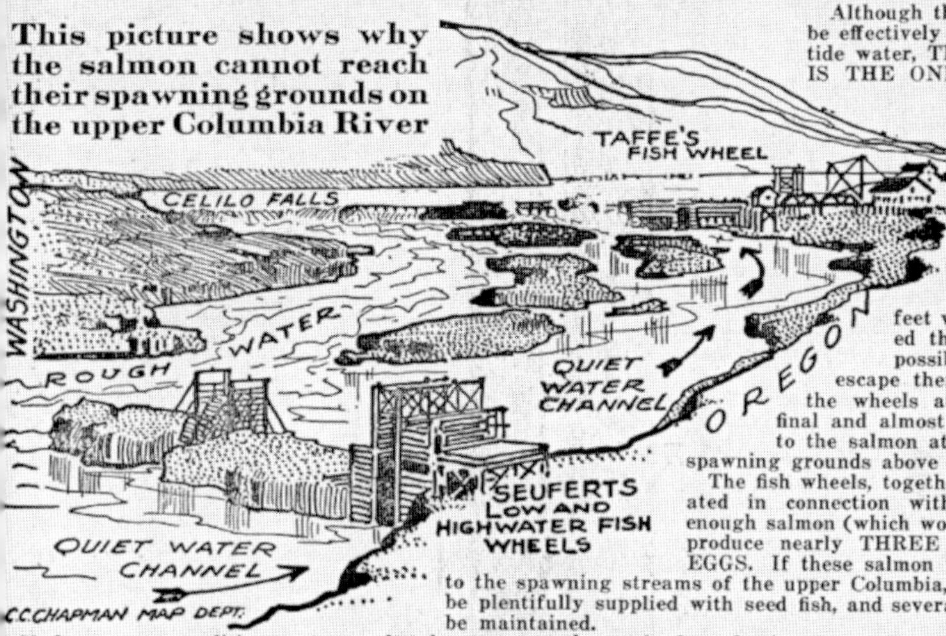

Although the deadly fish wheel can be effectively used on any river above tide water, THE COLUMBIA RIVER IS THE ONLY STREAM IN THE WORLD WHERE THIS MERCILESS GEAR IS PERMITTED.

They are located in the narrow gorge of the Columbia at Cascade Falls, The Dalles Rapids and Celilo Falls, where the river at some points is less than 200 feet wide. They are so situated that it is practically impossible for the salmon to escape them. This picture shows the wheels at Celilo, which form a final and almost insurmountable barrier to the salmon attempting to reach their spawning grounds above the falls.

The fish wheels, together with the seines operated in connection with them, take annually enough salmon (which would otherwise escape) to produce nearly THREE HUNDRED MILLION EGGS. If these salmon were allowed to escape to the spawning streams of the upper Columbia, these tributaries would be plentifully supplied with seed fish, and several large hatcheries could be maintained.

Under present conditions not enough salmon escape these wheels and seines to warrant the State in maintaining even an egg-taking station on these tributaries, which were once teeming with fish. Practically all of the Royal Chinook salmon eggs which are now used in the Columbia River Hatcheries must be taken from the Willamette river. The many millions eggs taken by two states and the Federal Bureau of Fisheries west of the Celilo Falls are of the fall chinook (or tullie) variety, which are of little value to the industry. This variety of fish this year sold on the market as low as two cents per pound.

FRANKLIN PRESS, ASTORIA, OREGON

(Please read other side)

As Columbia River salmon and steelhead returns from the ocean declined through the 1920s, public concern over the effects of commercial fishing increased. The Columbia River Compact enforced state harvest regulations, but commercial fish wheels continued to scoop millions of pounds of fish from the river. In 1914, the total salmon catch from the river was 38.5 million pounds. In 1915, it was 42.7 million pounds, and it stayed above 40 million pounds per year through 1919 before it began a steady decline. Only one year, 1925, was above 40 million pounds again. Fish wheels were seen as one of the primary problems, yet Oregon and Washington moved slowly, and at different times, to regulate them. Fish wheels were banned in Oregon in 1926, encouraged by public pleas like the one in this pamphlet, but not in Washington until 1935. (Courtesy Irene Martin.)

At Celilo Falls, the conflict of fishing cultures can be seen clearly in the c. 1895 photograph above, with a fish wheel (and outhouse) perched over the historic falls where tribal people had fished for centuries. Fixed gear like the fish trap below (and its scow to hold the fish, and the pick-up boat to pull the scow to the cannery), probably at Harrington Point in Grays Bay on the Columbia, could capture tens of thousands of pounds of fish in a day. Traps, also called pound nets or pond nets, were not outlawed in Oregon until 1949. In 1950, Oregon and Washington declared drift gillnets the only legal commercial fishing gear on the Columbia River for non-Indians. (Above, courtesy Old Oregon Photos; below, courtesy Washington State Archives.)

Drift gillnetting is an art, and the fishers are artists. They have to know not only what kind of net to use for different types and sizes of fish, but where to lay out their nets, how long to leave them, and when to gather them to check for fish. The first net fishermen on the Columbia used the power of sails and their own muscle to row out into the river, lay out their nets, and pull them in. This labor-intensive process got a boost in the early 1930s when gillnet boats were outfitted with gasoline engines. The addition of an engine also allowed boats to be fitted with power rollers in the bow—hence the name "bowpicker"—so the fisher did not have to pull the net in over the side. This invention also allowed fishers to work on their own without a crew, boosting their profits. Fisherman Kent Martin of Skamokawa, Washington, is pictured here in his bowpicker in 1964. His technique and equipment had not changed much over time. (Courtesy Irene Martin.)

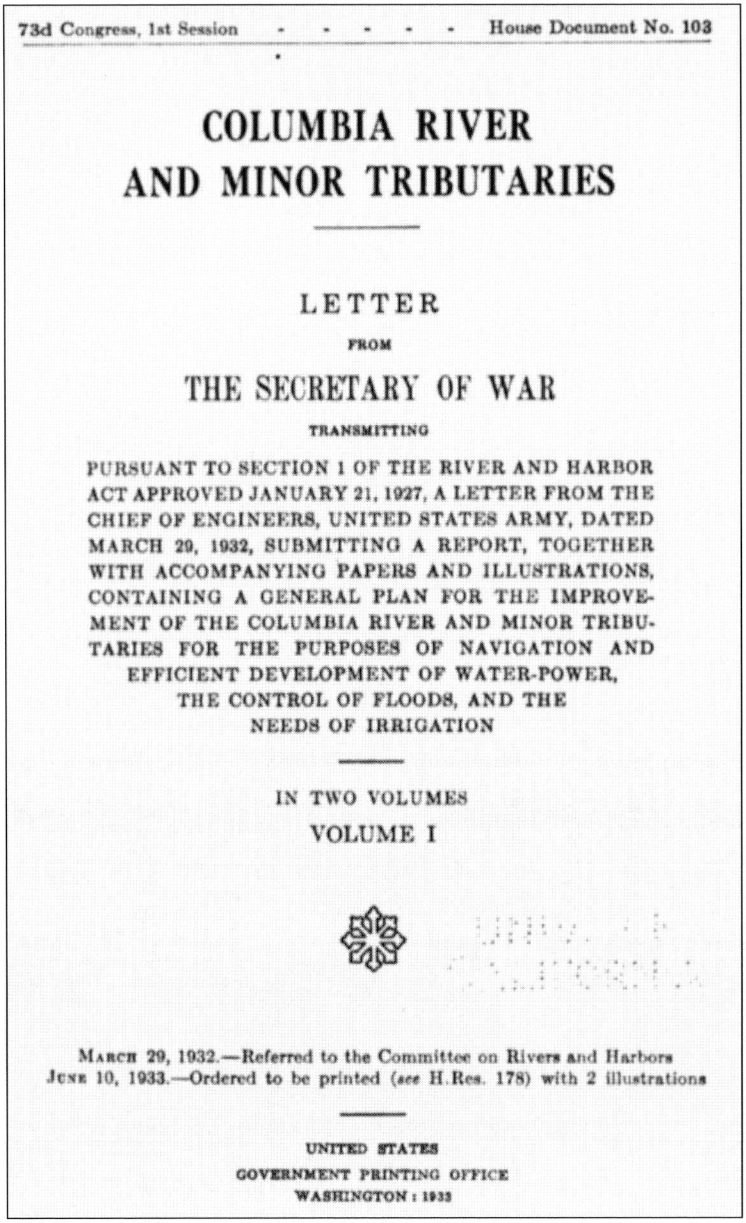

73d Congress, 1st Session - - - - - House Document No. 103

COLUMBIA RIVER
AND MINOR TRIBUTARIES

LETTER

FROM

THE SECRETARY OF WAR

TRANSMITTING

PURSUANT TO SECTION 1 OF THE RIVER AND HARBOR ACT APPROVED JANUARY 21, 1927, A LETTER FROM THE CHIEF OF ENGINEERS, UNITED STATES ARMY, DATED MARCH 29, 1932, SUBMITTING A REPORT, TOGETHER WITH ACCOMPANYING PAPERS AND ILLUSTRATIONS, CONTAINING A GENERAL PLAN FOR THE IMPROVEMENT OF THE COLUMBIA RIVER AND MINOR TRIBUTARIES FOR THE PURPOSES OF NAVIGATION AND EFFICIENT DEVELOPMENT OF WATER-POWER, THE CONTROL OF FLOODS, AND THE NEEDS OF IRRIGATION

IN TWO VOLUMES
VOLUME I

March 29, 1932.—Referred to the Committee on Rivers and Harbors
June 10, 1933.—Ordered to be printed (*see* H.Res. 178) with 2 illustrations

UNITED STATES
GOVERNMENT PRINTING OFFICE
WASHINGTON : 1933

Beginning in the late 1930s, hydropower was the primary instrument of economic change in the West, according to historians Dorothy Johansen and Charles Gates. Nowhere in the country is hydropower as important to the economy or the electricity supply as in the Pacific Northwest. The region relies on hydropower for more than half of its electricity, and most of it is generated at the 14 dams on the mainstem Columbia River and at others on tributaries, most of them federally owned. The Columbia is uniquely situated as a hydropower river, descending from 2,651 feet above sea level at Columbia Lake to the ocean in 1,248 miles, most of it through basalt canyons—ideal for anchoring big dams and generating big amounts of hydroelectricity. This vast hydropower potential initially was recognized in a study by the US Army Corps of Engineers and the Federal Power Commission published in April 1926. Subsequent surveys, presented in a report to Congress in March 1932, proposed eight dams on the Columbia, later increased to 10. (Author's collection.)

Five

The Hydropower Revolution

Franklin D. Roosevelt, campaigning for the presidency, greets a crowd in Portland on September 21, 1932. At a speech at the Civic Auditorium, Roosevelt made his vision for the Columbia clear. Its future, in his mind, was tied to hydropower. Roosevelt said, in part, "We have, as all of you in this section of the country know, the vast possibilities of power development on the Columbia River." (Courtesy the *Oregonian*.)

The first dam spanning the Columbia was Rock Island, seven miles south of Wenatchee, completed in 1933. Stone and Webster, an engineering firm that also managed electric utilities, built the 3,800-foot-long dam for the Washington Electric Company, a subsidiary of Puget Sound Power and Light. The 870-foot-long powerhouse could house 10 generators, but originally, only four were installed because there was no need for more power. (Courtesy Washington State Archives.)

Where could Roosevelt's dream and promise be fulfilled in the Columbia River Basin? The 1926 federal study pointed the way, and by the late 1920s, sites for the first two federal dams were confirmed: at a reach-of-tide location called Bonneville, 40 miles east of Portland and 146 miles from the ocean, and at Grand Coulee, 450 miles farther upriver. The Army report settled the Grand Coulee debate among pumpers and ditchers in favor of the pumpers. Grand Coulee Dam would raise the river 350 feet to create a lake, from which irrigation water could be pumped up to the Columbia Plateau. A site for Bonneville Dam was decided when two basalt uplifts were discovered under the river from shore to shore. (Above, courtesy US Bureau of Reclamation; below, courtesy US Army Corps of Engineers.)

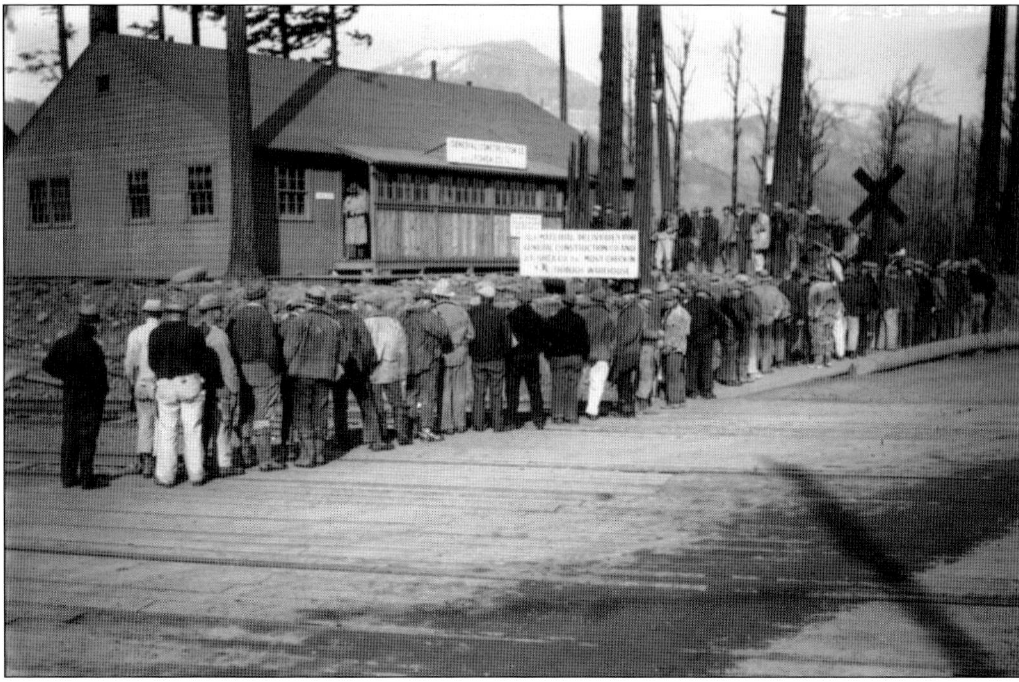

Bonneville and Grand Coulee Dams were enormous construction projects. It was 1935, in the depths of the Great Depression, and the projects employed thousands. Above, men line up to apply to work on Bonneville construction, which by then had been underway for two years. Bonneville, unlike Grand Coulee, would include a navigation lock to allow boats to pass, facilities to allow salmon and steelhead to pass upriver, and a reservoir stretching more than 100 miles upriver that would flood the Cascades rapids and make navigation easier. When construction began in 1933, Bonneville was a remote, unincorporated site along the railroad tracks and the Columbia River Highway with a tavern and campground that were popular with Portlanders. (Both, courtesy US Army Corps of Engineers.)

Each turbine runner in the dam weighed 900 tons and had a main shaft 40 inches in diameter. The runner is the "propeller" that turns with the force of falling water. As the runner spins, the shaft also spins. The spinning shaft is connected to the rotor, which holds electromagnets and spins within a stationary ring of coils called a stator. Electricity is drawn off the stator in three locations, each one a separate phase—hence three-phase, alternating current electricity. Each turbine unit could pass 13,000 cubic feet of water per second—enough to fill an average three-bedroom house. In these photographs, both taken in 1937, a group of workers pose on a turbine runner as it is installed in June (right), and a rotor is being installed in October. (Both, courtesy US Army Corps of Engineers.)

In August 1937, President Roosevelt dedicated Bonneville Dam, remarking that it would create "more wealth, better living, and greater happiness for our children." The first ships passed through the navigation lock in March 1938, and the dam officially opened four months later. It was a big project—too big for its critics. One derided it as the "Dam of Doubt" that would produce more electricity than anyone would need. But within a year, the Bonneville Power Project (today, the Bonneville Power Administration), signed its first power sales contract: with the Aluminum Company of America (Alcoa) for a huge smelter at Vancouver, Washington, 45 miles downriver. Two other Portland-area smelters followed, and the dam soon supplied power to build ships and airplanes during World War II. (Above, courtesy US Army Corps of Engineers; below, courtesy Bonneville Power Administration.)

Bonneville Dam officially opened on July 9, 1938. The dam had three fish ladders, water-filled staircases with each "step" one foot higher than the previous one. The dam also had two "fish lifts," like water-filled elevators. Over time, these did not work as well as the ladders and were abandoned. With the ladders, for the first time, the number of adult salmon and steelhead in the river could be counted as they swam upriver past the dam. The 1938 total was 469,027 fish, primarily Chinook (271,799), but also steelhead, sockeye, and coho. Fish passage was always part of the dam plans, but ladders as big as those needed at Bonneville—a vertical climb of 72 feet—had never been built anywhere in the world. (Both, courtesy US Army Corps of Engineers.)

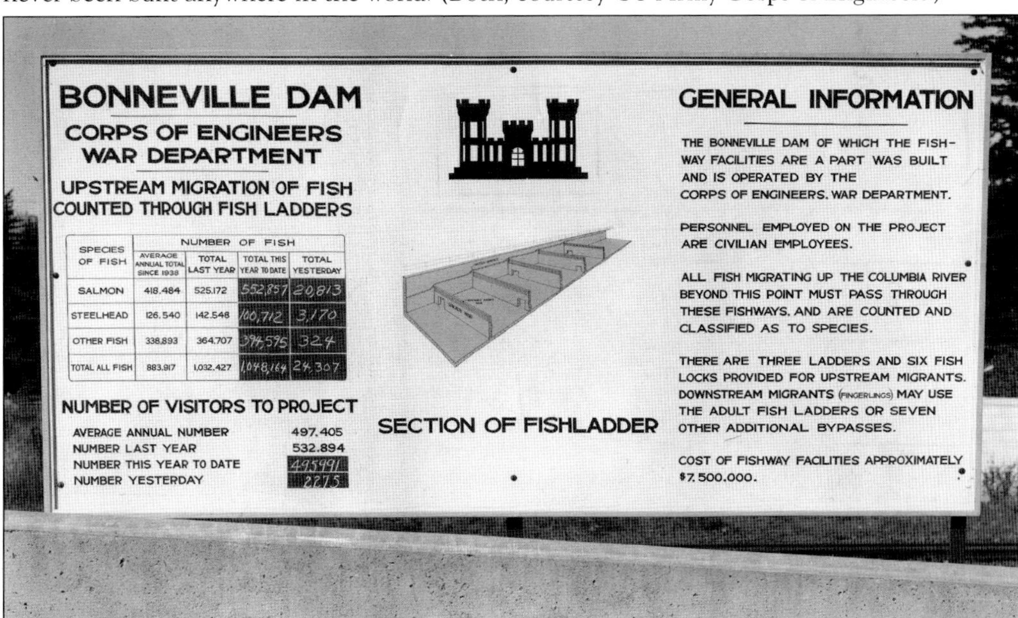

Meanwhile, in the semi-arid Columbia Plateau of central Washington, irrigation was the cause that united agriculture and other interests, beginning around 1920, behind the idea of building a great dam at Grand Coulee and pumping water from its reservoir up to the plateau. In the 1910 photograph above, a man poses on horseback on the plateau, an area known for its sagebrush and often-failed attempts at dryland farming. At left are two views of the canyon where Grand Coulee Dam would be built, and Steamboat Rock, which today overlooks the irrigation reservoir Banks Lake. (Both, courtesy Washington State Archives.)

On September 9, 1933, exactly four days after the official start of construction at Grand Coulee, engineers drilling test holes found bedrock and set the stakes that marked the center line of the dam. Above, surveyors pause for a photograph at the Grand Coulee site in 1933. The image below shows the initial work at Grand Coulee. In March 1934, Congress passed the Fish and Wildlife Coordination Act (amended in 1946 and 1958), the beginning of efforts to mitigate the inevitable impact of Grand Coulee and other federal Columbia River Basin dams on fish and wildlife. (Both, courtesy Bureau of Reclamation.)

Grand Coulee might never have been the dam it is if not for the tireless work of a small group of true believers in irrigation and economic development, led by the indomitable Rufus Woods, owner, editor, and publisher of the *Wenatchee Daily World* newspaper. Woods had a lot of fun at the *World*. He liked to dress in goofy clothes and drive around in a convertible that featured a portable desk and typewriter, always ready to report the news. Woods firmly believed the great dam would be as vital as food and water. Picking up on an idea of his friend Billy Clapp, an attorney in nearby Ephrata, Woods trumpeted the idea in the *World* on July 18, 1918, under the page-one, all-capitals, banner headline, "Formulate Brand New Idea for Irrigation Grant, Adams, Franklin Counties, Covering Million Acres or More." It was not the first proposal for a dam at the Grand Coulee of the Columbia, but the story energized and excited people about the dam's potential. (Courtesy the *Wenatchee World*.)

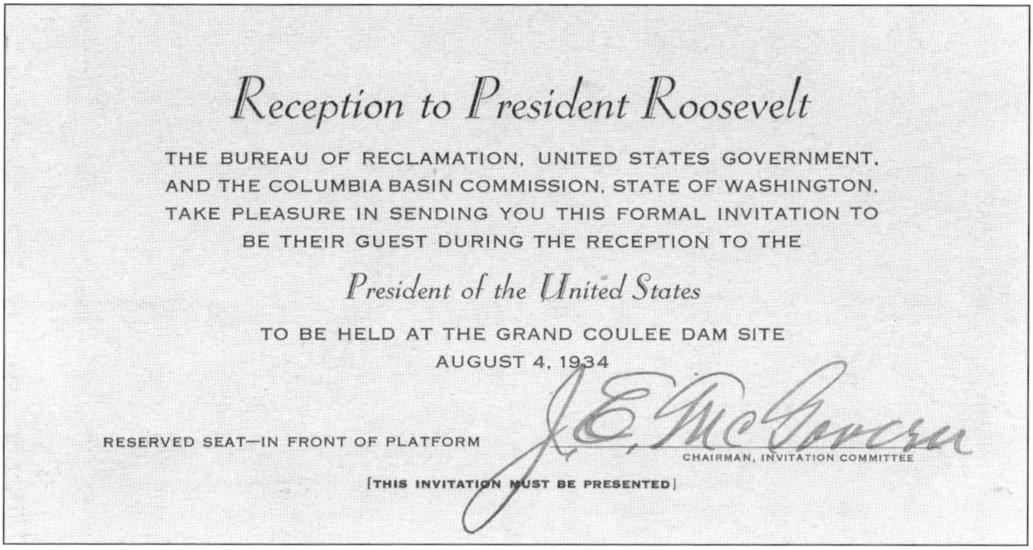

Reception to President Roosevelt

THE BUREAU OF RECLAMATION, UNITED STATES GOVERNMENT, AND THE COLUMBIA BASIN COMMISSION, STATE OF WASHINGTON, TAKE PLEASURE IN SENDING YOU THIS FORMAL INVITATION TO BE THEIR GUEST DURING THE RECEPTION TO THE

President of the United States

TO BE HELD AT THE GRAND COULEE DAM SITE
AUGUST 4, 1934

RESERVED SEAT—IN FRONT OF PLATFORM

CHAIRMAN, INVITATION COMMITTEE

[THIS INVITATION MUST BE PRESENTED]

President Roosevelt was the hero of Grand Coulee Dam, at least at the federal level. On August 4, 1934, with dam construction underway for a year, Roosevelt spoke at an event at the site celebrating the beginning of construction. The size of the crowd, estimated at 20,000, astonished the federal officials traveling with the president. Just a day earlier, Roosevelt spoke at the Bonneville Dam construction site, also to a large crowd. Historian Paul Pitzer notes in his history of the dam a remark attributed to Eleanor Roosevelt as she looked out at the mostly treeless, sun-blasted basalt and sagebrush landscape around Grand Coulee: "It was a good salesman who sold this to Franklin." (Both, courtesy Northwest Room, Spokane Public Library.)

Into the rolling hills punctuated by basalt outcroppings, grassy hills, and isolated ranches about 80 miles northwest of Spokane came thousands of workers hoping to escape the economic chaos of the Great Depression and build better lives for themselves by building the great dam. Their sheer numbers overwhelmed the limited housing available in the area, and so temporary housing was built, including a floating camp. More than 12,000 people worked on the dam between 1933 and 1943—as many as 7,400 at any one time. They came from all 48 states; more than 60 percent were between 20 and 29, most were single, and about 40 percent worked as unskilled laborers, writes Pitzer. These photographs show construction camps downstream from the dam site (above) and workers lining up for their pay (left). (Both, courtesy Washington State University Libraries archives.)

It was dangerous work. As cofferdams were built in the river to create dry places for the concrete and steel, those who worked over water wore life jackets. The project was enormous and complex. In 1939, the year these photographs were taken, construction progressed the fastest and farthest, with crews working around the clock. More than 2,000 men worked each shift, and at night, floodlights gave the scene the look of a science fiction movie. Ultimately, the dam would be 450–500 feet thick at its base, 30 feet thick at the top, and contain 11,975,521 cubic yards of concrete, three times as much as Boulder (Hoover) Dam on the Colorado River. (Right, courtesy Washington State University Libraries; below, courtesy US Bureau of Reclamation.)

The construction of Grand Coulee Dam at river mile 596 eliminated runs of anadromous fish from the river above that point. Before the dam, fish spawned in the river and in tributaries all the way to the headwaters at Columbia Lake, and so Grand Coulee cut off fish from more than 650 miles of the mainstem river and thousands of miles of tributary spawning habitat. The government knew this would happen and promised to build hatcheries for salmon and steelhead downriver and for resident fish species like trout and kokanee in the reservoir behind the dam to compensate, but the hatcheries never produced the numbers and genetic diversity of the populations that were lost. While the dam was under construction and before it rose high enough to create an impassable barrier, a temporary fish ladder was installed, pictured here in 1938, the last year of its operation. Until 1943, salmon continued to bump their heads against the concrete of Grand Coulee, trying to get upriver. (Courtesy Washington State University Libraries.)

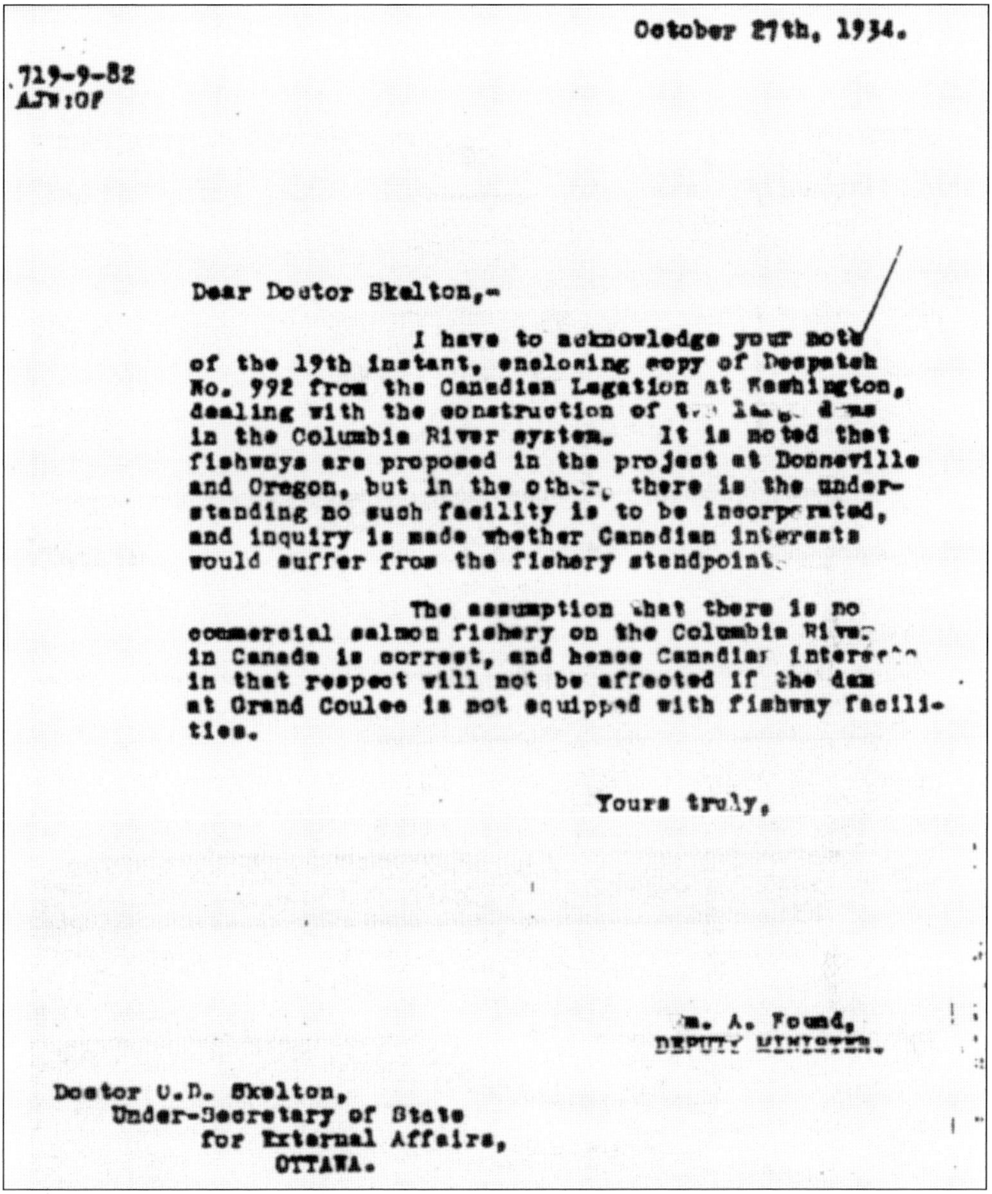

October 27th, 1934.

719-9-82
AJF:OF

Dear Doctor Skelton,-

I have to acknowledge your note of the 19th instant, enclosing copy of Despatch No. 992 from the Canadian Legation at Washington, dealing with the construction of the large dams in the Columbia River system. It is noted that fishways are proposed in the project at Bonneville and Oregon, but in the others, there is the understanding no such facility is to be incorporated, and inquiry is made whether Canadian interests would suffer from the fishery standpoint.

The assumption that there is no commercial salmon fishery on the Columbia River in Canada is correct, and hence Canadian interests in that respect will not be affected if the dam at Grand Coulee is not equipped with fishway facilities.

Yours truly,

Wm. A. Found,
DEPUTY MINISTER.

Doctor O.D. Skelton,
Under-Secretary of State
for External Affairs,
OTTAWA.

Meanwhile, the Canadian government appeared truly indifferent to the fish loss Grand Coulee would cause. Only a few local citizen groups expressed concern at hearings on the dam's impact —on the river level at the international border, not on salmon—convened by the International Joint Commission in Spokane and Trail, British Columbia, in 1941 (ironically, the year the dam was completed). Neither British Columbian nor Canadian federal fishery agencies testified. Canadian federal officials were not entirely silent, though. In October 1934, the Canadian Legation in Washington, DC, took note that building the dam without fish passage meant ruin for Canadian Columbia River salmon. The legation expressed concern in a letter to the federal Ministry of Fisheries in Ottawa. The reply, signed by a deputy minister, concluded, "The assumption that there is no commercial salmon fishery on the Columbia River in Canada is correct, and hence Canadian interests in that respect will not be affected." Remarkably, the Canadian government wrote off Columbia River salmon because they were not fished commercially in the river. (Author's collection.)

There was a rescue effort, though. In January 1938, the Washington Department of Fisheries issued a detailed report on how to save the salmon and steelhead runs. The report proposed trapping adult fish at Rock Island Dam and transporting them in tank trucks, like the one pictured here at Bonneville Dam, to release locations in rivers and streams both above and below Grand Coulee and eventually to a hatchery that would be built at Leavenworth, on a Wenatchee River tributary called Icicle Creek. From there, juvenile fish would be transported to release sites throughout north-central Washington, in the hope that they would adapt to a new environment. It was an ambitious plan, but it lacked public support. Enthusiasm for the dam was greater. The annual trapping and hauling began in May 1939, and by 1943, the progeny of those fish began returning to spawn. Returns were a fraction of the pre-dam numbers, and never have been since, but nonetheless, the Reclamation Service claimed success. (Courtesy US Army Corps of Engineers.)

North of the border, First Nations people felt the fish loss acutely. The Sinixt, Okanagan, Ktunaxa, and Secwepemc had fished for salmon in their homelands for millennia. Biologists estimate that more than 85 percent of the sockeye crossing Rock Island Dam once spawned in rivers and lakes above Grand Coulee. Many headed to Arrow and Slocan Lakes in Canada, part of the Sinixt homeland encompassing present-day northeastern Washington and southeastern British Columbia. In *The Geography of Memory: Recovering Stories of a Landscape's First People*, author Eileen Delehanty Pearkes spoke to Sinixt elder Eva Orr. Asked to identify her favorite traditional food, Orr does not hesitate: salmon, from the sea. Delehanty Pearkes wrote, "Eva Orr's ironic answer refers to a rich heritage now all but forgotten, a time . . . she was just old enough to have experienced first-hand, before dams severed . . . the Interior's connection with a distant ocean." Today, the Sinixt and other tribes from Canada and the United States are working to reintroduce salmon to abundant coldwater habitat above Grand Coulee Dam. Pictured here is a tribal fisherman at Kettle Falls. (Courtesy Northwest Room, Spokane Public Library.)

The huge turbines for Grand Coulee, built on the East Coast, were so big and heavy that a special route had to be mapped for the trains that would carry them to avoid low underpasses, tunnels, and narrow bridges. The trip from Pittsburgh for the first generator took 11 days, and all its parts required 38 flatcars. Shown above being assembled in July 1941, the first generator, labeled L3 (third position, left powerhouse looking downriver), was the first 108,000-kilowatt generator in the world. The completed dam, seen below in June 1942, is 550 feet tall from bedrock to the top, contains 11 million cubic yards of concrete, is about 4,000 feet across, and cost $162.6 million to build—about $16 million under budget. (Above, courtesy Washington State University Libraries; below, courtesy Bureau of Reclamation.)

On June 14–16, 1940, the Confederated Tribes of the Colville Reservation hosted a three-day gathering to eulogize the loss of the historic Kettle Falls to rising water behind Grand Coulee. Appropriately, the event was called the "Ceremony of Tears." US senator Clarence Dill, a longtime supporter of Grand Coulee, was a featured speaker at the event on June 16. Dill acknowledged the terrible impact the dam would have on Indians by wiping out their historic fishery, but also said he hoped that the Indians would realize some benefit from the power the dam would produce. Leaders of the San Poil, Colville, and other tribes also spoke. A reporter for the *Spokesman-Review* newspaper of Spokane estimated the attendance at 1,000 and also wrote that there was a "more serious side" of the event at which the Indians "told of their sadness of the passing of the falls, and some thought the government should reimburse them for their loss." (Courtesy Northwest Museum of Arts and Culture, L85-143.380.)

On March 22, 1941, a crowd of about 8,000 gathered at the dam for speeches and the ceremonial beginning of power generation by one of the two small turbines that would provide power to the dam and also to the town of Coulee Dam and sections of Mason City. The celebration included a parade led by the Grand Coulee High School marching band. Chief Jim James of the Nez Perce Tribe, accompanied by other tribal members, pushed a button that set the turbine running and then gave a speech in his native language. At the insistence of news reporters, area tribal leaders also posed for photographs. It was a day of celebration for non-Indians at the dam, but a day of immense sadness for James and other tribal members who could hardly celebrate the end of the salmon and steelhead runs that were crucial to their culture, economy, and sustenance. (Courtesy Bureau of Reclamation.)

The loss of fish and fishing in British Columbia was felt most acutely by the First Nations people, for whom salmon was important culturally and nutritionally, but it was also a blow to non-tribal communities. Fish were an important source of protein, and the loss of salmon put a dent in the food supply as well as in recreational fishing. These photographs show fishing success in the Arrow Lakes in the 1930s. (Both, courtesy Arrow Lakes Historical Society.)

Grand Coulee was part of a New Deal push to electrify rural areas with public power. In this photograph, members of the Adrian Grange near Soap Lake, Washington, rally in support of publicly owned (consumer-elected) utilities, an effort that led to the formation of the Grant County Public Utility District in 1938. In 1941, the Bonneville Power Administration decided to feature the newly completed Grand Coulee in a revision of its 1939 film *Hydro*. Stephen Kahn of Bonneville's Information Office hired Woody Guthrie, a folk singer who had appeared on radio programs in New York and Los Angeles, to write songs for the film and potentially appear as an actor. Guthrie was hired from May 13 to June 13, 1941, during which time he wrote 26 songs. Some, such as "Roll on Columbia," are sung to this day. World War II halted the production of the film; it was finally completed in 1949 using several Guthrie songs. (Courtesy Grant County Public Utility District.)

In 1943, the Bonneville Power Administration received an urgent request for a large block of electricity to serve a secret load along the Columbia River between Richland and Yakima. The load was for the classified Manhattan Project nuclear reactors that would be built on the river near the small community of Hanford to make plutonium for atomic weapons. The plutonium in the bomb dropped on Nagasaki was manufactured at Hanford. Above, the Hanford B Reactor is seen between the water towers on the right. Meanwhile, work on the Columbia Basin Project to irrigate the Columbia Plateau region south of Grand Coulee slowed during World War II, but picked up again after the war. Water from Lake Roosevelt behind the dam is pumped up to the project reservoir, Banks Lake. (Above, courtesy US Department of Energy; below, courtesy Library of Congress.)

Irrigation brought the fertile but arid soils of the mid-Columbia River Basin to life. The federal government sponsored irrigation projects in the Yakima River Basin, a Columbia tributary in central Washington, and on the Columbia Plateau above Grand Coulee Dam. Hops, pictured above being strung in the Yakima project in 1936, were a popular crop. Water flowed into the canals of the Columbia Basin Project for the first time on May 29, 1952. To celebrate, boosters encouraged World War II veterans to come to the area to farm. To raise awareness, they sponsored a "Farm in a Day" giveaway. Volunteers built an entire farm, including a home, furnishings, outbuildings, and newly planted crops, in 24 hours. The recipient was a 30-year-old veteran from Kansas. (Above, courtesy Library of Congress; below, courtesy US Bureau of Reclamation.)

Industries that relied on the Columbia River boomed during and after World War II, from canning fish in Astoria to making aluminum at the Alcoa smelter and building Liberty ships at the Kaiser Shipyard, both in Vancouver, Washington. Alcoa was the first of 10 aluminum smelters in the Northwest, all taking advantage of inexpensive hydropower from Columbia River dams. The Alcoa smelter opened in early 1940, a year and a half after the completion of Bonneville Dam. The dam provided electricity for the energy-intensive smelter. The Vancouver shipyard, which opened in early 1942, turned out more than 140 ships and two drydocks during World War II. (Both, courtesy Bonneville Power Administration.)

A lifestyle, a historic fishery, a falls—all were radically changing as the Columbia was steadily tamed behind dams and put to work generating hydroelectricity and providing for irrigation, inland navigation, and flood control. As the steamboat era slowly came to a close and giant dams created reservoirs, the fishing lifestyle shown at left was interrupted. Below, surveyors are measuring the area around Kettle Falls to determine how far up the shoreline the reservoir rising behind Grand Coulee would go—what would be flooded and what would remain above the water. Railroad tracks, farms, and homes would have to be destroyed or moved out of the way of the impoundment. (Left, courtesy Washington State University Libraries; below, courtesy Washington State Archives.)

Six

Dams, Flooding, and International Cooperation

Competition with other types of transportation brought an end to the steamboat era on the Columbia. In the United States, the last run was in 1947 by the *Georgie Burton*; in British Columbia, the *Minto* made its last run in 1954. This photograph shows, from left to right, the *Columbia*, *Minto*, and *Bonnington* tied up at Nakusp in the early 1940s. (Courtesy Arrow Lakes Historical Society.)

In late May and early June 1948, a Columbia River flood destroyed the city of Vanport, Oregon. Vanport was a temporary city built in 1942 to provide housing for wartime shipyard workers and their families who arrived from all over the country. The racially diverse population swelled to as much as 40,000, making Vanport the second-largest city in Oregon behind its neighbor Portland. By the time of the flood, May 30, 1948, Memorial Day, the population was about 20,000, and the housing was deteriorating. On that day, with the Columbia running about eight feet above flood stage, a railroad dike failed and floodwaters poured into Vanport. The flood overturned houses—they had been built quickly and without foundations—and killed a total of 32 people. In the aftermath, Congress authorized more water-storage dams upriver in the Columbia River Basin. At the time, there were only three dams on the Columbia—Bonneville, Rock Island, and Grand Coulee—and only Grand Coulee had flood control as one of its authorized purposes. (Courtesy Oregon Historical Society.)

In 1946, Congress authorized construction of Chief Joseph Dam 51 miles downstream from Grand Coulee; it began operation, without fish passage, in 1955–1958. Chief Joseph is the upriver extent of salmon and steelhead migration in the Columbia, blocking access to about 700 miles of the mainstem and all the associated tributaries where salmon historically spawned. A 1938 report by the US Army included a plan to build four dams to provide slackwater for navigation along the lower Snake River from its confluence with the Columbia to Lewiston, Idaho, a distance of 140 miles. Four dams were completed between 1961 and 1975: Ice Harbor (below), Lower Monumental, Little Goose, and Lower Granite. (Above, courtesy Washington State Archives; below, courtesy US Army Corps of Engineers.)

In 1947, the Columbia Basin Interagency Committee, which coordinated federal interagency and state interests in Columbia River development, conducted a public hearing on the question of whether to support a moratorium on construction of the four lower Snake River dams until there was some assurance that the dams would have effective passage facilities for salmon and steelhead. The Corps of Engineers was ready to move ahead with construction, but many fish scientists were concerned about the effectiveness of fish passage at the then nine-year-old Bonneville Dam downriver, pictured here. The position of the federal government was clear, though. Warner W. Gardner, assistant secretary of the interior, wrote to the committee that "the salmon run must, if necessary, be sacrificed," adding that "the government's efforts should be directed toward ameliorating the impact of this development upon the injured interests and not toward a vain attempt to hold still the hands of the clock." (Courtesy US Army Corps of Engineers.)

The Northwest's population grew by 44 percent from 1940 to 1949. Net immigration to the Northwest was more than one million. In 1953, Pres. Dwight Eisenhower appointed George Boldt, a Tacoma, Washington, judge, to the US District Court, Western Washington District. Twenty-one years later, in 1974, Boldt authored one of the most important legal decisions in the history of Northwest salmon recovery efforts. Also in 1953, Eisenhower shifted the nation's power policy to encourage local utilities to build dams on major rivers. Three public utility districts in central Washington, aided by investor-owned utilities, took advantage of this shift and built four dams on the Columbia during the 1950s and 1960s: Priest Rapids (originally authorized as a federal dam) and Wanapum, built by the Grant County Public Utility District; Rocky Reach, built by the Chelan County Public Utility District, which by now also owned Rock Island Dam; and Wells, built by the Douglas County Public Utility District. Priest Rapids, pictured here under construction, began operation in 1958. (Courtesy Washington State Archives.)

The Dalles Dam, a federal project constructed by the US Army Corps of Engineers, began operation in 1957. The photograph at left shows blasting during construction. The Dalles Dam is at the city of The Dalles, Oregon, at river mile 191.5. On March 10, 1957, the flood gates closed, and the reservoir behind the dam quickly filled, drowning the historic Indian fishery at Celilo Falls. The loss of the fishery, pictured below in 1942, was a devastating blow to Columbia River tribes, who had fished at the falls for millennia. (Left, courtesy Rick Schafer/Ray Atkeson Image Archive; below, courtesy Old Oregon Photos.)

In 1944, the United States and Canada asked the International Joint Commission (IJC) to investigate the potential of the transboundary Columbia for "a greater use than is now being made." In 1959, the IJC reported on how hydropower and flood-control benefits might be shared by the two countries. Negotiations culminated in the Columbia River Treaty in 1961, ratified three years later. This 1964 illustration published in the *Oregon Journal* of Portland captures the goodwill and collaboration codified in the treaty, but some Canadian residents and politicians called it a "giveaway." The treaty mandated the construction of three dams in British Columbia (Keenleyside, Duncan, and Mica) for flood control and optimum hydropower generation downriver in the United States. In return, British Columbia receives a portion of the additional hydropower. The treaty also authorized the option to construct Libby Dam on the Kootenai River, a Columbia tributary, in Montana 48 miles south of the border for flood control and power. All four dams were built. Lake Koocanusa, Libby's reservoir, flooded farms and ranches 42 miles into British Columbia. (Courtesy the *Oregonian*.)

The treaty has no expiration date and will continue indefinitely unless one country requests termination or renegotiation with 10 years advance notice. In 2013, each country announced an intention to renegotiate, not terminate, and negotiations have been conducted periodically since. In the treaty preamble, the two countries state that construction of three dams in British Columbia would provide benefits in the form of additional hydroelectricity and flood control "in a manner that will make the largest contribution to the economic progress of both countries and to the welfare of their peoples of which those resources are capable." Here, High Arrow Dam, later renamed Hugh Keenleyside, is under construction in 1966. This dam flooded the third-largest agricultural valley in British Columbia at the time, destroyed Sinixt archaeological sites, and forced the resettlement of farms, homes, and communities along the river. (Courtesy Arrow Lakes Historical Society.)

The treaty preamble also noted that the people of the United States and Canada "have, for many generations, lived together and cooperated with one another in many aspects of their national enterprises for the greater wealth and happiness of their respective nations." The treaty was seen as a natural evolution of that cooperation, but in the places where the treaty dams would create reservoirs in British Columbia, people and communities had to move. Productive farmland was lost, forests were cut down, some structures were burned, and others moved to higher ground. Homes also were moved to make way for Lake Roosevelt behind Grand Coulee Dam, but the impact was less intense, overall. In British Columbia, the places lost were more populated and productive. (Both, courtesy Arrow Lakes Historical Society.)

Above, an Anglican church arrives at Nakusp after being floated up the Arrow Lakes. Below, the Beaton, British Columbia, school is moved to higher ground. The construction of Keenleyside Dam raised the water level about 40 feet. The provincial hydropower authority, BC Hydro, paid compensation to people who lost their homes and income, but many considered it inadequate, and the grand words of cooperation and friendship in the treaty rang hollow. Adequate compensation was not achieved until 1995 with the creation of the Columbia Basin Trust. Over time, the trust became the major force in mitigating the effects of the treaty dams on the economy, environment, and societal well-being of this area of the province. (Both, courtesy Arrow Lakes Historical Society.)

The march of hydropower continued through the 1960s. In 1967, the Nathaniel Washington Power Plant, also called "the Third Powerplant," and the largest at Grand Coulee, was completed, made possible by the water control created by the Columbia River Treaty dams upriver. Its total capacity is 4,215 megawatts—two-thirds of the total capacity of Grand Coulee. Two full-size buses could be parked side by side in any of its six penstocks, each of which is capable of carrying an amount of water equal to nearly twice the average annual flow of the Colorado River. In 1968, John Day Dam on the Columbia at river mile 215.6 was completed. The three treaty dams in British Columbia were finished between 1968 and 1973; two are on the Columbia. Today, there are 14 Columbia River mainstem dams from Bonneville at river mile 146 to Mica, one of the treaty dams, at river mile 1,018, which is 200 miles north of the headwaters. (Courtesy of Bureau of Reclamation.)

By 1960, the population of the four Northwest states was 5,964,000, an increase of 884,000 since 1950 and 2,058,000 since 1940. The Columbia now provided millions of acre-feet of water annually for irrigation. Irrigated acres grew from 3.5 million in 1928 to 6.5 million in 1966. Salmon and steelhead continued to be overfished in the lower Columbia, and fish counts at Bonneville Dam declined. In 1950, a total of 717,500 fish were harvested between the dam and the ocean, and 157,200 fish were harvested upriver from the dam. Ten years later, the harvest was 492,100 downriver from the dam and 3,900 upriver. Between 1961 and 1964, annual production at Columbia River salmon canneries, most of them on the river downstream from Bonneville, dropped to its lowest level since about 1870. (Courtesy Northwest Power and Conservation Council.)

Between 1961 and 1967, several large hydroelectric dams began operation. In 1961, Ice Harbor Dam at Snake River mile 9.7 came online, providing power to the Bonneville Power Administration grid and also to the Hanford Nuclear Reservation. Also in 1961, Rocky Reach Dam came online at river mile 473.7 on the Columbia near Wenatchee. The dam was built by a consortium led by the Chelan County Public Utility District. One of the partners was Alcoa, which built a smelter at Wenatchee to use power from the dam. Downriver, the Grant County Public Utility District put Wanapum Dam in operation in 1963 at river mile 415.8. In 1967, the Douglas County Public Utility District put Wells Dam into operation at river mile 515.1. (Above, courtesy US Army Corps of Engineers; below, courtesy Douglas County Public Utility District.)

In a landmark ruling in 1974, US District Court judge George Boldt ruled that Indians had a right to half of the harvestable surplus of salmon and steelhead in Northwest rivers based on language in treaties the tribes signed with the United States in 1855. Four years later, the 50-percent issue reached the US Supreme Court in *Washington v. Washington State Commercial Passenger Fishing Vessel Association*. The Supreme Court rejected Washington's argument that Indians and non-Indians had an "equal opportunity" to the fishery, stating, "In our view, the purpose and language of the treaties are unambiguous; they secure the Indians' right to take a share of each run of fish that passes through tribal fishing areas." The court further held that the harvestable portion of each run that passes a tribal fishing area should be divided "approximately equally" into treaty and non-treaty shares. (Above, courtesy Northwest Power and Conservation Council; below, courtesy Columbia River Inter-Tribal Fish Commission.)

In 1975, the last of four federal dams built on the lower Snake River, Lower Granite, began operation. The dams, built for both navigation and hydropower, made Lewiston, Idaho, 465 miles from the Pacific, a seaport. The next year, in 1976, Congress authorized the Lower Snake River Compensation Plan to mitigate the impacts of the dams on salmon and steelhead by constructing hatcheries and improving spawning and rearing habitat. The plan includes 10 hatcheries and 16 satellite facilities on the river and tributaries in Idaho, Oregon, and Washington. (Courtesy US Army Corps of Engineers.)

In the mid-1970s, the Bonneville Power Administration and the Washington Public Power Supply System (WPPSS) began construction of five nuclear power plants to supplement power from the federal Columbia River hydroelectric power system, in response to forecasted growing demand for power and future power shortages. But the forecast was wrong, and four plants were never completed, including plant three, pictured above. Bonneville, obligated to pay for plants one, two, and three, raised its rates several hundred percent in the early 1980s, and its customer utilities in turn raised rates to their ratepayers. The effect, as illustrated below, was shocking. The debt for plants four and five, backed by public utilities, was canceled by the Washington Supreme Court. In all, the debt for the five plants totaled $24 billion. (Above, courtesy Washington State Archives; below, courtesy Northwest Power and Conservation Council.)

In December 1980, after four years of work on legislation to respond to the WPPSS crisis and its effect on utilities and ratepayers, and also in recognition of declining salmon and steelhead runs, Congress approved and Pres. Jimmy Carter signed into law the Northwest Power Act. The act authorized the four Northwest states of Idaho, Montana, Oregon, and Washington to form the Northwest Power and Conservation Council (the agency was known until 2003 as the Northwest Power Planning Council) as an interstate compact. The act gave the council responsibilities for mitigating the effects of Columbia River Basin hydropower dams on fish and wildlife; long-term regional power planning giving priority to energy conservation to meet future demand; and informing and involving the public. The council met for the first time in April 1981. The first council was made up of, from left to right, (first row) Al Hampson, Oregon, Dan Evans, Washington, Bob Saxvik, Idaho, and Keith Colbo, Montana; (second row) Roy Hemmingway, Oregon, Chuck Collins, Washington; Chris Carlson, Idaho; and Gerald Mueller, Montana. (Courtesy Northwest Power and Conservation Council.)

The 1980s were significant for recreation, fishing, and salmon protection in the Columbia and Snake Rivers. The first houseboats were launched on Lake Roosevelt in 1987; the previous year was the first of eight years of below-average precipitation in the basin, which resulted in reduced hydropower generation. In 1986, Congress authorized the creation of the 90-mile Columbia River Gorge National Scenic Area, which stretches from about Washougal, Washington, to Rufus, Oregon, imposing land-use restrictions to protect scenic values and limiting development. In 1988, the three mid-Columbia public utility districts—Grant, Chelan, and Douglas—agreed to operate their five Columbia River dams in a manner that protects wild fall Chinook salmon at Vernita Bar downriver of Priest Rapids Dam (above), at the head of the 50-mile Hanford Reach (below), the last free-flowing section of the river in the United States. (Above, courtesy Grant County Public Utility District; below, courtesy US Department of Energy.)

On April 9, 1990, the Shoshone-Bannock Tribes of the Fort Hall Indian Reservation, Idaho, filed a petition with the National Marine Fisheries Service to list Snake River sockeye salmon, like these pictured in Redfish Lake, Idaho, as endangered under the Endangered Species Act. Two months later, the Fisheries Service received another petition, this one from multiple environmental groups, to list Snake River spring, summer, and fall Chinook, and lower Columbia coho salmon as threatened species. All of the species eventually were listed, leading to a succession of lawsuits against the government that continue to this day alleging inadequate protection of the listed species. (Both, courtesy Idaho Department of Fish and Game.)

On August 10, 1996, fast-growing trees in a filbert orchard in the Portland area shorted a major transmission line, initiating a cascade of power outages as far south as Mexico, affecting four million customers in eight states. Most power was restored in about eight hours, but the event focused public attention on the electricity interdependence between the Pacific Northwest and Southwest and the importance of Columbia River hydropower to the entire West Coast. Power from the four federal dams on the lower Snake River is part of that system, but the dams have been a lightning rod for controversy over their impacts on salmon and steelhead. Some say they should be removed; others say they should stay. Pictured below is Little Goose Dam. (Left, courtesy Northwest Power and Conservation Council; below, author's collection.)

In February 2014, Columbia Basin tribes from the United States and First Nations in British Columbia issued a report on the potential to reintroduce salmon and steelhead to the Columbia River Basin above Chief Joseph and Grand Coulee Dams, including into British Columbia. Since then, Upper Columbia tribes have identified suitable habitat above the dams to support salmon and steelhead once they are reintroduced. The tribes also have experimented with juvenile and adult salmon releases into the river above the two dams using surplus hatchery fish from downriver. These releases are celebrations for the tribes. Here, Rodney Cawston and his family release an adult salmon from a transfer bag into Lake Roosevelt behind Grand Coulee in 2019. (Courtesy Confederated Tribes of the Colville Reservation.)

Warm water kills coldwater species like salmon and steelhead and encourages the growth of warm-water species, some of them invasive, like Northern Pike. In June and July 2015, parts of the river warmed to lethal levels. A quarter-million sockeye, like this one showing signs of heat stress, died in the 146 miles between Bonneville and McNary Dams. Northern Pike, a nuisance species, invaded Lake Roosevelt behind Grand Coulee Dam in about 2009, probably as the result of an illegal introduction upriver, and continued to move downriver as they multiplied. The Spokane, Colville, and Kalispel tribes have aggressive programs including gillnetting and electrofishing to remove these predators. (Courtesy US Geological Survey Western Fisheries Research Center.)

In 1938, Lewis Mumford, an American historian and philosopher, visited the Northwest, including the Columbia River Gorge. He fell in love. In his 1939 follow-up report, Mumford enthused about the "breath-taking landscapes," and "the subtle and manifold beauties of the Columbia River Gorge, which unrolls itself like some great kakemono of classic Chinese landscape art." Despite intense development across the Columbia River Basin, today, there are still places of magnificent beauty. This tension between the natural and the developed environment is key to understanding the Columbia. In his book *Northwest Passage*, author and historian William Dietrich predicts "constant challenge, and a daunting, promising, contentious, exhilarating future" for the river, in which "we continually remake our relationship to nature." (Courtesy Kate Hotchkiss.)

Bibliography

Allen, John Elliot, and Marjorie Burns, with Sam C. Sargent. *Cataclysms on the Columbia: A Layman's Guide to the Features Produced by the Catastrophic Bretz Floods in the Pacific Northwest.* Portland, OR: Timber Press, 1986.

Delehanty Pearkes, Eileen. *The Geography of Memory: Recovering Stories of a Landscape's First People.* Nelson, BC: Kutenai House Press, 2002.

———. *The Geography of Memory: Recovering Stories of a Landscape's First People.* New ed. Victoria, BC: Rocky Mountain Books, 2022.

———. *A River Captured.* Victoria, BC: Rocky Mountain Books, 2016, 2022.

Dietrich, William. *Northwest Passage: The Great Columbia River.* New York, NY: Simon & Schuster, 1995.

———. *Northwest Passage: The Great Columbia River.* 2nd ed. Seattle, WA: University of Washington Press, 2016.

Johansen, Dorothy O., and Charles M. Gates. *Empire of the Columbia.* New York, NY: Alfred A. Knopf, 1957.

Lichatowich, Jim. *Salmon Without Rivers: A History of the Pacific Northwest Salmon Crisis.* Washington, DC: Island Press, 1999.

Martin, Irene. *Legacy and Testament: The Story of Columbia River Gillnetters.* Pullman, WA: Washington State University Press, 1994.

Nisbet, Jack. *Sources of the River: Tracking David Thompson Across Western North America.* Seattle, WA: Sasquatch Books, 1994.

Pitzer, Paul. *Grand Coulee: Harnessing a Dream.* Pullman, WA: Washington State University Press, 1995.

Scholz, Allan. *Fishes of Eastern Washington: A Natural History, Volume 1.* Cheney, WA: Eastern Washington University Biology Faculty Publications, 2014.

Seufert, Francis, with Thomas Vaughn. *Wheels of Fortune.* Portland, OR: Oregon Historical Society, 1980.

White, Richard. *The Organic Machine: The Remaking of the Columbia River.* New York, NY: Hill and Wang, 1995.

Willingham, William F. *Water Power in the 'Wilderness:' The History of Bonneville Lock and Dam.* Portland, OR: US Army Corps of Engineers, 1997.

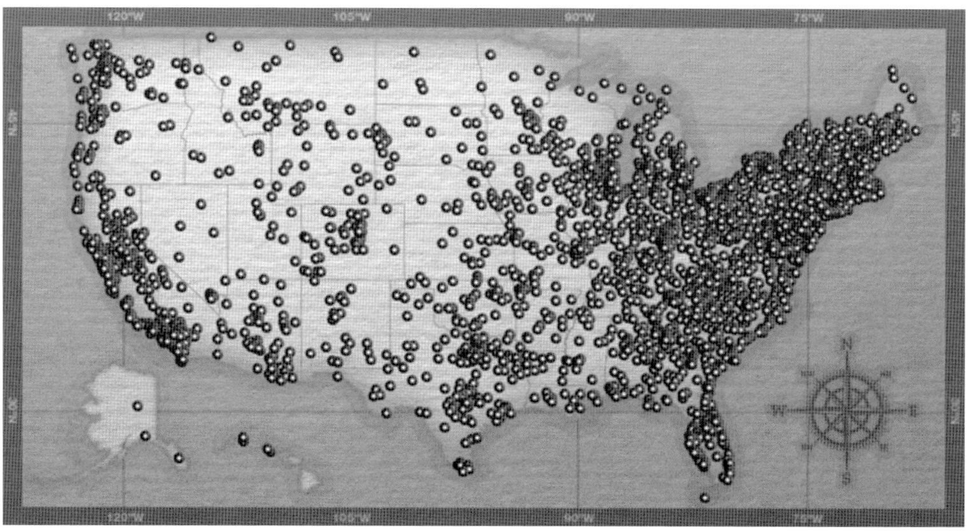